Creation or Evolution

Does it Matter?

By

ROBERT W. RIDLON, JR.

and

ELIZABETH J. RIDLON

ISBN: 1-4033-4434-5 (e-book)
ISBN: 1-4033-4435-3 (Paperback)
ISBN: 1-4033-4436-1 (Hardcover)

Library of Congress Control Number: 2002108207

This book is printed on acid free paper.

Printed in the United States of America
Bloomington, IN

Scripture refrences are from a New International Version (NIV) and are copied from the macBible™ Software for Bible Study and Research. © 1990, Zondervan Electronic Publishing, Grandrapids.

The transmission electron micrograph of the cell and all photographs were taken by the authors.

1stBooks - rev. 03/09/04

Contents

Preface

Creation or Evolution: Does It Matter? is a book of hope. The debate over two diametrically opposed explanations for the origin and diversity of life is relentless and enduring. The stakes are high because ultimately the issue is people and their reason for existence. The debate will not be won in a courtroom or on a college campus or at a Sunday School class — it will be won in the hearts and minds of individuals. Until then, we live with a persistent undercurrent of evolutionary thought that explains our existence as accidental. This implies living a life without an ultimate purpose and may explain the continuous decay in the world's value on human life.

Alternatively, the view that we were created by a loving God, gives purpose and meaning and hope. This is a God that is in control and has a plan — a plan that includes people.

In this book, we survey the beginnings of evolutionary thought; the people, the motives, the methods, the conclusions, the problems, and the unanswered questions. The premises of evolution are established as: Life originated by random chance events and diversity (species) is due to environmental pressures favoring certain variations. We examine the modern view of Darwinian evolution and the current models used to explain origin and diversity.

We use the Biblical scriptures to outline and describe the events that establish the premise for the creation model: All life was created by God. We attempt to provide a framework within which a scientist can operate in seeking to better understand the world and its inhabitants. Creation Science, as a discipline, enforces scientific observation, analysis, and conclusion for seeking the truth about the origin and diversity of life.

Robert W. Ridlon, Jr.
May 2002

Part One: It's About People

CHAPTER ONE

Does it Matter Where We Came From?

As a family, we traveled extensively for over 20 years. It's just one of those things we liked to do. Even today, we never tire of seeing new places and meeting people from different places around the world. Regardless of where we were, Cairo, Egypt or the Florida Everglades, there was often that first question: "Where are you from?" I'm not sure why that question is asked, but we ask the same thing when we meet someone new. What's the first question you usually ask a new student in your class at school? What's one of the first questions you ask the new co-worker? If you think about it, it is probably "where are you from?"

Where we are from is a big part of our identity. Everyone is from somewhere. It may be a place that they are glad to have behind them or it may be a place that brings back good memories.

Have you noticed the elaborate measures many adopted children take to find their biological parents? A small industry has even developed which assists in the process of reuniting children with their biological parents. Even when the children were raised in the happiest of adoptive homes and never knew their biological parents, there is sometimes a deep need for knowing where they came from. Sometimes it is a happy reunion and sometimes it doesn't work out. But, knowing often brings needed closure in this area.

These examples are in a way superficial examples of a bigger question: Where did we ultimately come from? You exist, your friends and co-workers exist, historical figures existed before us; but, where did we all come from? Even if you think space travelers brought us here from a distant galaxy, they had to come from somewhere. Therefore we are faced with the question and we need the answer. There can only be two possibilities. Let's see what they are and how they may influence our feelings and actions.

One possibility is that through some series of events over time, involving the random movement of chemical components in a primordial environment, a meaningful set of replicating components arose which is life itself.

With that in mind, let's say your little eight year old daughter crawls up on your lap and says, "daddy, where did we come from?"

In terms she would understand, you could answer, "Millions and millions of years ago, there was just a lot of mud and slime around and that was it. One day something in the slime bumped into something else in the slime and made a better slime. All this mud and slime began to make bigger and better things until — well, here we are."

Does this sound encouraging? Does this even sound reasonable? Would this help the little eight year old girl want to live a life that counts? In a couple years (or less) someone will ask that little girl if she wants to try some drugs. If she thinks she is from slime, she may not care what kind of slime she takes into her body. Possibly, that's how people do see it and that's why they go on living life like it was some sort of slime accident. Next time you are on a new job or are a new member of a class, when someone asks where you're from, you could say, "I am from slime, and proud of it."

The slime story would be ridiculous and laughable if it wasn't the sad explanation that evolutionists want everyone to believe. So, what's the alternative? The other possibility is that we were carefully and wonderfully created and made by a loving God for a special purpose. Let's see how that works with the little girl on her daddy's lap.

"Daddy, where did we come from?"

"A long time ago, God made everything in the world and even in space. He made people to be very special so they could love Him and He could love them back and take care of them forever."

Now the little girl goes off to play or to school and when someone offers her some drugs she can think perhaps for a moment to herself, "I am special — God made me," and have a reason to say no.

Knowing Our Future

Our origin can and does impact the image we have of ourselves. We are all very much like that impressionable little eight-year-old girl. If you believe you arose from slime, then that may affect the decisions you make about how you treat your body from moment to moment and day to day. In other words, think about the kind of self-image you will have based on evolution's notion that we have somehow plodded along a million year course as a result of a series of accidents beginning with some early primordial slime. On the other hand, by considering that we are here because of the intentional act of a God that loves us, we see value in ourselves because God values us. What a difference that makes in our present decisions.

Now let's for a moment consider the impact of our origin on our future as well. What does the person who has subscribed to the slime theory (evolution) have to look forward to? A better job? Marrying another person who is from the slime? Maybe someday dying and turning into the slime from where they came? Doesn't sound like a very happy outlook to me. Guess what — it isn't a very happy outlook. In fact, it looks sort of hopeless, does it not? Could the excesses we see in our world be an expression of hopelessness? When we think of excesses, we often think first of the United States and not the less affluent third world nations. However, there are multitudes of people in those less privileged countries that are engaged in excesses, just not the materialistic kind. They may be involved in drugs and illicit sex, for example. In the United States, we are no better. We have perhaps some additional excesses attributable to our freedoms and higher standards of living. Where's the hope in the drunkenness and sex pervasive on our school campuses? Where is the hope in the high divorce rate? Where is the hope in teen suicide rate, the crime rate, pornography, and out of control drug abuse? The answer is, there is no hope reflected in this type of lifestyle. This is a lifestyle that is lived out day after day by people who have no hope of a future because they have no basis for hope. After all, they believe they are here because of an accident, so they may tend to live their lives like accidents. In their minds, there is no future beyond what they can realize in the moment, and even that stinks for the person whose origin is the slime of evolution. We ought to have great compassion for these people. They are blind to the truth and desperately missing out on the hope that could make all the difference.

Alternatively, let's consider the implications of knowing that our origin is from God. That means much in terms of our future as well. If we were created by a God who loves us, then we ought to expect that God will take care of us as well. The same Bible that explains our origins, explains our future. We can know God and trust that He has our best interests in mind. We can count on the promise of our future in Heaven. We can be sure that when we leave our present physical bodies, that we will forever be in the loving presence of the God that created the universe (and us). Of course this requires us to fully accept and trust God by faith. What a difference that can and does make in the lives of people. If this were a lie or wishful thinking, then there would be no hope. However, we have many evidences that point us to God. One of these is the general revelation of God that is exhibited in His creative work — that is the creation. This is often expressed by reference to these two scriptures:

"The heavens are telling of the glory of God; and their expanse is declaring the works of His hands" (Psalm 19:1).

"For since the creation of the world His invisible attributes, His eternal power and divine nature, have been clearly seen, being understood through what has been made, so that they are without excuse" (Romans 1:20).

A Big Decision

I don't think there is much argument about which of the two possible explanations makes a nicer story. The first alternative (evolution) really doesn't sound too good. The second alternative (creation) sounds a lot nicer. But, which is true? We will look at both evolution and creation and examine the evidence and explanations of both. Both are really models that attempt to explain life in two diametrically opposing ways.

The arguments go beyond mere science and logic. The case for creation is strong and consistent with what is observed in nature and what the Bible says. When the Biblical account of the origin and diversity of life is used as the model, the things we observe scientifically make sense. It is easy to see the science within the boundaries of the scripture — everything fits together. There is not only meaning to life, there is an understandable relationship between the components of nature. The facts we can observe in nature all point to the Biblical model. It however acknowledges a God for which we are accountable; but, a God that loves us and created us for a purpose.

The case for evolution is remarkably weak scientifically. I know that sounds contrary to what you may have learned or thought. Hopefully, as you read this book, the weaknesses, mistakes, and wrong answers of evolution will become more apparent. There is nothing wrong with science. Science is a wonderful set of disciplines that help us see and understand things. Like any toolset, science can be misused and therefore yield wrong outcomes. When the predictions and premises of the early evolutionary thinkers are examined and the lack of evidence is revealed, the decision becomes an informed one.

CHAPTER TWO

God's Plan

In the beginning, God had a plan. He doesn't have a Plan-A and a Plan-B, and he doesn't make it up as he goes. God still has the same plan — His plan. Genesis (the book of beginnings) records the first events of a plan that is still unfolding. It's important to understand a few things about God in order to understand Genesis and the rest of the Bible and the rest of the Plan.

Who and What is God?

God is eternal; he has always existed. That may be incomprehensible due to our limited thinking. However, it is an apprehensible concept — that is, we can accept it without fully understanding it. God's eternal existence is a key component to understanding and embracing the true meaning of creation because it requires a Creator.

God has always been God and he will always be God. There may be a struggle between good and evil among the earth's inhabitants, but it isn't because God is struggling with evil. God allows the evil one (Satan) to have certain influences over things, but Satan is defeated. The world as we know it will run its course in God's timing and the victory will become known to all.

Many false religious systems in the world are based on a dualistic idea whereby the gods are in a constant battle with each other. When the "good" gods are winning, there is prosperity in the nation or in the household. When the "bad" gods are winning, there is famine and hardship among the practitioners. Much of the religious practice in this system (*i.e.,* sacrifice and ritual) is an attempt to please, stimulate, or cheer-on the right god, thus trying to optimize the chances of good things happening.

The concept of good gods versus bad gods is not the case for the true God and the true believers of Christianity. The true God cannot and will not be defeated. God is not dependent upon anyone for support or help. Whether we love God or not — He is still God. This means God is in control and only allows or causes things to happen that are within His will and plan.

God does know everything — all things of the past, present, and even the future. Not only did He know what He was going to do regarding the creation, He also knew that it would fall apart due to the sinful nature of man. This often troubles people. Why would God create something that He would eventually have to destroy? That's a question that may only be answered by God Himself. There are however, some things to consider. First, we know that God knows what He is doing. He isn't at all surprised when things go wrong. When God created man, he did so with the purpose of creating beings that would fellowship and communicate with Him. If God had created man in the way we create robots or computer programs, he could have encoded this creation with some really neat words and phrases like *please, thank you, I love you*, etc. My computer can say *I love you*, but I know that it was only my program that directs the statement — the computer has no choice. So, since God wanted our love for Him to be meaningful, He provided us with a choice. Giving a choice brought the opportunity to rebel against God by the very first man and woman on earth in the Garden of Eden. Every person that has ever lived has the same opportunity. Therefore when we embrace and love God, it is meaningful because we have a choice. It is also meaningful to understand that God loved us first.

God even knows us personally and has known us before we were born. Listen to what it says in the Bible about that:

O Lord, you have searched me and you know me. You know when I sit and when I rise; you perceive my thoughts from afar. You discern my going out and my lying down; you are familiar with all my ways. (Psalm 139: 1-3)

All the days ordained for me were written in your book before one of them came to be (Psalm 139: 16)

Problems and Solutions

God knew that fellowship with mankind could and would be broken. Remember, after Adam and Eve ate from the tree that God had forbidden, they hid from God. The fellowship was broken. Adam and Eve even sewed fig leaves to cover themselves, but when they heard the sound of God, they hid themselves. This symbolizes the inadequacy of man's abilities to provide a way of reconciliation and restoration. When Adam and Eve broke fellowship, it was God that provided the way for restoration by providing a covering of clothes made of animal skins. Providing the skins of animals required the shedding of innocent blood.

The subsequent corruption that occurred after Adam and Eve were expelled from the Garden of Eden began with their son Cain's murder of his brother Abel. Cain was evil and rebellious and angry. Several generations passed from that time and there was apparently an excess of evil among mankind which required God to dismantle the creation with devastation by a global catastrophic flood which washed clean the earth. It's important to note here that God wasn't surprised by the corruption; however, he was disappointed. (Gen 6:5-6)

The concept of animal sacrifice was apparently perpetuated. When Noah and his family departed the ark, Noah's first order of business was to build an altar and offer burnt offerings. This system of sacrifice was practiced in a prescriptive way throughout the period of the Old Testament. These sacrifices are reminiscent of the first shedding of innocent blood by God Himself in the garden of Eden when He provided the skin coverings for Adam and Eve. The sacrifices themselves were never intended to pay the price required to restore the fellowship. That came later when Jesus voluntarily laid His life down as the innocent blood shed for us.

Here are some characteristics of obedient sacrifice in the Old Testament period:

- Had to be repeated.
- Expressed man's acknowledgement of God's way as the only way.
- Served as a reminder that remission of sins required the shedding of innocent blood.
- Provided a covering for sin so fellowship with God could be restored.
- Served as a model for the ultimate once for all sacrifice provided by God — Jesus.

The important point to comprehend here is that God deeply loves us and desires fellowship with us. That is why we were created in the first place. Even our rebellion is dealt with. First, by a system of sacrifices for the faithful to practice demonstrating their faith in God; and second, looking forward to His provision of a redeeming sacrifice someday. It's important to point out here that it wasn't the physical act of sacrificing animals that accomplished anything. It was the repentant and obedient heart of the individual that was reflected in the act of sacrifice.

The Old Testament believers could count on God's future promise. After Jesus came fulfilling that promise, believers can look back on that event as a promise fulfilled and accept that way of restoration. That's why we don't need sacrifices anymore. The once for all promised sacrifice was made by Jesus on the cross. He was the innocent one whose blood was shed for us.

The Big Picture

Looking back is sometimes a good way to see the future. Looking back over the Bible beginning in Genesis, we can see the stories of real people in real places, dealing with real problems, and the presence of a real God.

God's Plan began with the creation story in Genesis and goes on to describe the forefathers that led the way in the unfolding events of history. Incidentally, people and events of the Bible have been the subject of criticism over hundreds of years. For example, obscure nations, such as the Hittites, were often dismissed as fictional adding to the attacks of skepticism. In God's good timing, these criticisms have melted away with discovery after discovery by archaeologists and other scholars. Today, nearly every turn of the archaeologist's shovel adds confidence to the veracity of the Bible.

As the pages of the Old Testament are turned the nature of mankind — a nature of selfish rebellion, is revealed. However, we see a God who remains unchanged in His love for the people. The Bible describes the details of mankind and its relationship to this loving God. God says to trust Him alone, yet many chose other gods and their own way. The Bible records a patient and merciful God who seeks to restore the rebellious, time after time.

God gave the Law, through Moses, as a way to reveal His nature, to guide the people into a godly lifestyle, and also to show us our sinfulness. Any infraction of the Law is falling short of the measurement God would apply. However, remember, God provided the way of restoration when man failed — the obedient sacrifice that covered the transgression and foreshadowed the coming of a savior that would pay their debt in full. The Old Testament is a foundational framework of covenants and promises that God would keep ultimately through the cross when Jesus was crucified to pay for those sins committed by the Old Testament believers.

Besides the historical accounts, the Old Testament part of the Bible contains some beautiful literature and poetry. These books reveal God's nature and the kind of relationship that is pleasing in God's eyes and His will.

Some of the most poignant books of the Old Testament are the 17 prophetic books. The writers of these books were specially appointed as God's spokesmen. They acted on God's behalf to bring God's words to the people in order to bring them to godliness and holiness. They spoke words from God that revealed God's future plans and God's warning about continued rebelliousness. They proclaimed that there would ultimately be a final judgement that would come at the end of the age and all sin would be judged. This meant that God's people would be saved and the unbelievers would be judged and punished. The prophets also spoke of the coming Savior.

The Savior Jesus came as promised and as the prophets said. Beginning in the New Testament gospels, we can read about the birth of Jesus, His life, and ministry. He ultimately came to lay His own life down as the once for all sacrifice. This was the once for all payment of the sins of the people who lived in the Old Testament times — that is the ones who, by faith, acknowledged God and His ways and realized they must seek Him only. You might say that the Old Testament believers had racked up a pretty big debt over the years and Jesus paid it at the cross.

In the New Testament period, after His death and resurrection, people still have a debt of sin that needs to be paid. We can look back to the cross (so to speak) and realize that our debt has been paid as well. The New Testament books of the Bible are divinely inspired, but humanly written, beginning with four Gospel accounts of the life of Jesus, His ministry on earth, His death on the cross, and His resurrection from death. After these four Gospels, there are books that tell of the early Christian church, how Christianity spread, and encouragement to live the victorious Christian life. The Bible ends with the book of Revelation which promises a blessing on all who read it and take it to heart. This final book is the testimony of Jesus regarding the ultimate victory over sin and death.

What About You?

Let's face the facts. We will only be staying on earth for a very short period of time. It can possibly be 70 - 100 years before we die; but, sooner or later we must go. What lies ahead — beyond the grave?

God's desire is for us to have eternal life with him. That's why we were created. You can have that eternal life beginning now. God's Big Plan includes you.

> For God so loved the world that he gave his one and only Son, that whoever believes in him shall not perish but have eternal life. (John 3:16)

If you would like to know more about God's love for you, there is a short discussion of this in Appendix F.

As we will see in the upcoming chapters, there is a definite structure to the earth and all the living things — especially man.

CHAPTER THREE

Seeing People Through God's Eyes

We stepped down from the platform as some 300 college students applauded. Walking across to the adjacent room to sign books, we both felt a sense of self-satisfaction. The lecture on creation science was given nearly perfectly (as we saw it). Many times in recent months, we illustrated the beauty and logic of creation and dismissed the evolution model premise by premise. There was a special sense of satisfaction at this particular event, having dealt with a few physics students on the issues of radiometric dating. Having spent some long hours as science students ourselves and recently considering uranium decay, we were armed with a quick and deadly response to their assertions that U-238 decay was a valid accurate way to reconcile the age of certain geologic formations. Notice the word *deadly*. There had been a particular group of students, probably three or four, that had formed a small coalition in a dimly lit section at the back of the room. Were they there to challenge our assertions of creation by a Creator? Were they there out of curiosity, seeking to hear more about a model that challenged their training and notions? Did it even matter why they came?

We sat congenially in the little room set aside for us to meet and talk informally with students after our formal presentation. We answered questions and signed copies of our two books on creation science. We even entertained more debate on a few esoteric topics like speciation and natural selection. All seemed well as we packed up and started our walk down the dark street to our car that fall evening. A staff member from the host organization offered to carry some of our equipment to the car for us and we accepted his offer. As we walked along, we were expecting the usual comments like "Thanks." or "That was great." But, there was only silence — one of those

14

pregnant silences when one feels there was something that was waiting to come forth — yet didn't. We said thanks for the help with our stuff and still expected a comment in return, but there was none. What did this mean?

Our presentation went well enough. In the preceding months, we had given essentially the same talk to home school groups, churches, youth groups, and college students delivering the compelling truth about creation and evolution. Why this feeling of coldness this time? Why not the usual euphoria? The evolutionists were turned to dust and for the creationists in the audience, it was music to their ears. Upon arriving home the next day, we sent an email to the host group asking specifically for their reaction. The first paragraph in their answer included the words that we wanted to hear: *excellent, helpful, appreciated*. The second paragraph, however, contained some words we didn't want to hear: This is from the second paragraph of that email:

> "…many students involved in the ministry mentioned that they desired for an evolutionist to come away from the night not feeling attacked. They believed that some did because of jokes made about the model of evolution. A first [time visitor] might have felt that we were all about giving truth but not about giving grace."

That hit us hard. Who did they think they were telling us how to give a talk? After all, aren't the creationists the ones that are always the brunt of the jokes? Aren't we the ones that are considered stupid and ignorant of science? Despite the estimates that there are thousands of scientists who are Christians and creationists, we are the ones being picked on. Besides, the evolution model is funny and pointing out its errors and even making jokes about some of the goofy stuff people believe about it is always fun.

Reflecting back on the talk, was the approach justified? Opening comments explained how believing in evolution made random chance events the reason for our existence and that believing that way was like saying that we exist because of an accident. If one thinks we exist because of an accident of nature, we will live our lives like an accident — devoid of hope and purpose. On the other hand, if we understand that we were intentionally created by a God who loves us, then we will live our lives with hope and with a purpose. We are always careful not to identify with so many axe grinders that seem to put creation science on a pedestal as some sort of religion or worse — lose sight of the Creator. The creation model is presented against the backdrop of the loving God of the universe. So why the perception of antagonism? There may be a clue in some things said in previous paragraphs — perhaps something in the attitude.

Remember about providing a deadly response to the arguments about uranium decay. Deadly? That's a pretty serious word to use. Deadly is defined as "causing or tending to cause death." Was that in the message? Was this causing death to the argument or "killing" the one arguing? What about the later thought of turning the evolutionist to "dust"? That seems more directed to the person not the argument. Even though we think we know what dust is, the dictionary is helpful in making my point. One definition is: "something of no worth." Wait a minute. We are the good guys who start the talk by saying that we are created by a loving God with a purpose, right? How does turning an evolutionist to dust play into that? It doesn't. Was this a message of compassion, having the Truth and wanting to share it; or a barrage of criticism hurled from a defensive crouch?

Whether we are arguing against abortion or evolution or heavy metal music, we must take care not to turn the objects of our love (those that don't see the Truth) into objects of hate. Believe me, we never thought of anyone in that audience as an object of hate. Their spears of confrontation were amusing at worst. We do have a genuine love for the lost person and understand that condition can cloud their understanding of Truth. However, care must be taken to not reduce that person to a state of "dust" lest the message of love and truth be cloaked.

We learned a valuable lesson that evening. It was a humbling one. If we are to be effective in our representation of the Truth, we must do so in a manner that reflects the love of the God of Truth who first loved us. "While we were yet sinners..." Whether dealing with junior high, the young folks in high school and college, or the professionals in our community, we must see them through God's eyes. That means seeing them as the objects of God's love.

Conclusion

Viewing the lost and confused person differently may not seem to have an effect on their ability to deal us some serious antagonism, but it can change the way we respond to it. Do we give up the fight? No! Do we become weak and acquiescent to the world's twisted view? No! Do we compromise the Truth and become accepting of wrong ideas and behaviors? Never!

So how can we deal with the lost person (or the misguided ones) in a gentle, consistent, loving, yet uncompromising way in the areas of creation and evolution? If we "lose" an argument, what are the stakes? Are we even supposed to "win" an argument, or should we be focused on presenting the truth?

The following chapters will attempt to answer these questions and equip the Christian to confidently and lovingly address the ever-present important questions of our origin. We will examine the premises of the creation model, as it is presented in the Bible, and examine the evolution model's explanation as well. By looking at the science (methodical observations, analysis, and conclusions) of the observable world, we will develop a strong almost irrefutable case for creation and show the vast weakness and improbability of evolution.

We will learn to see our audience as "honored guests" invited to partake in a meal of sweet Truth. Whether they like what we serve is not up to us. We can only do our best to be God's ambassadors. I am reminded of Paul's guidance to Timothy:

> And the Lord's servant must not quarrel; instead, he must be kind to everyone, able to teach, not resentful. (2Tim. 2:24)

CHAPTER FOUR

The Emperor's New Clothes

Various polls in recent years reveal a surprisingly high number of Americans (as high as 50%) believe in creation, in some form, as opposed to evolution. This may seem high considering the relentless powerful indoctrination that prevails in the public school system, colleges, magazines, and newspapers, providing continuous "evidence" for evolution. Nonetheless, there is a solid substantial resistance to accepting evolution and its hopeless conclusions.

Before we explore this further, let's understand what the creationist's believe in contrast to what the evolutionist's believe about life.

Creation Versus Evolution

There are two main principles that can be derived from the creation model. The first is that life (all life) was created through an intentional act of God — not by chance. Second, the diversity (or differences) in species is attributable to God's original creation — God's original plan. Although, we see variability within a species (*e.g.,* dogs can be poodles, dalmatians, chihuahuas, collies, etc), dogs won't evolve into bears and dogs didn't come from frogs. These species remain distinct. We do see variation and hybridization take place under forced protocols. One example is selective breeding of cattle to provide better milk producers. For years and years, grain hybridization has been underway to produce corn, wheat, beans, and other crops that resist drought and disease. The 140 or 150 different breeds of dog are another common example of how variation is possible within a species. This is no contradiction to creation; it is testimony of how God provided a capability for organisms in a

19

population to vary in their morphology (or physical form) to accommodate changes in the environment and become the dominant variation. They don't become new species, however. In biology, this is called natural selection which correctly might be thought of as a God-given repertoire of genetic potential — pretty smart! These basic premises are not only observed as true, they are found in Genesis Chapters 1 through 9, as well as other Biblical references (listed in Appendix E).

Evolution, on the other hand, explains the origin of life as a random chance event that involved non-living material becoming living. Life was generated by some set of circumstances, which includes the random assemblage of particles. The second premise is that over time (millions and millions of years) mutations occurred in these organisms which amounted to an improvement that defined those organisms as a successful new species. The idea of a Creator (God) is dismissed and instead it suggests that life was essentially an accident. These premises are not observed today. Life cannot be created in a laboratory setting and we do not see one species evolving into another.

We will spend much more time evaluating these two models in upcoming chapters. Now that we know a little more about these diametrically opposing ideas, let's get back to the question of why the growing rejection of evolution.

The Emperor Gets New Clothes (or Not)

Besides reading the polls about Americans in general, it is particularly encouraging that there are also a growing number of scientists that embrace creation and creation science as well. If you read the biographies of some of the world's greatest scientists in history, you will be surprised (or not) at the testimonies. Well known among these are Leonardo da Vinci, Blaise Pascal, Isaac Newton,

Charles Babbage, and Louis Pasteur. Were these brilliant men unenlightened by the modern biochemistry of today, or did they know the truth?

In the early nineteenth century, a young Danish man named Hans Christian Anderson wrote some very interesting, thought-provoking, and funny tales as well as some serious novels. Among these were some well-known stories such as *The Little Mermaid* and *The Ugly Duckling*. One little story that stands out is *The Emperor's New Clothes*. The following is a brief review of this story, but it would be worth it to get a hold of a copy and read it first hand for yourself. It is truly a work of genius that was actually written before Darwin and his theory of evolution.

In the story, the emperor of the city was preoccupied with new clothes to the extent that he was willing to spend all of his resources on them. One day some unscrupulous men showed up and claimed to be able to make the most beautiful clothes imaginable. In fact, the tailors said that the special clothes were invisible to the stupid and incompetent. Only the worthy people would be able to see the clothes. Of course the king liked this idea. He would not only get the fine clothes he wanted, but he had the added opportunity to differentiate between the smart and the dumb people. The emperor gave the would-be clothes-makers a large sum of money and the process was underway. Everyone in the town knew of the clothes and how the stupid people would not be able to see the fine clothes. The emperor sent one of his trusted courtiers to see how things were going. As expected, the courtier saw nothing, but pretended to see the clothes in all their glory and splendor. The emperor sent another courtier to check progress and, like the first, he saw nothing as well; but, likewise he commended the quality and beauty of the goods. Finally, the emperor himself, along with a number of his court, went out see the loom. Of course neither he nor any of his court saw anything. The emperor wandered if he was himself stupid after seeing nothing. He nonetheless pretended to see the clothes and extended his hearty approval of its quality and beauty. The emperor even named the weavers of the "fake" clothes as "Imperial Court Weavers." The emperor put on his new clothes and marched in a procession through the streets. In order not to appear stupid, everyone called the clothes beautiful, but, in fact they couldn't see the clothes — the clothes didn't exist. Finally, an innocent child proclaimed, "But he has nothing on at all." Eventually all the people began to repeat the child's words, "But he has nothing on at all." The emperor heard what they had reasoned, but didn't relent. The story ends with the emperor continuing the procession. The final sentence in the story says "And the chamberlains walked with still greater dignity, as if they carried the train which did not exist."[1]

The Emperor's New Clothes and Evolution

Evolutionary theory is in many ways just like this story of the Emperor and his hopes and obsession with the special clothes. There are some obvious parallels between the events of this story and the story of evolution theory. There are some remarkable similarities between the characters in the Andersen story and the characters that have shaped evolutionary thought. What is even more interesting is the various responses exhibited once the child had exposed the truth. Eventually all the people saw it, but for some — those entrenched in the folly of their endeavor, there was no turning back.

In much the same way today, we are often told that unless you accept and believe in the "obvious" fact of evolution, then you are stupid. In the scientific community this is even more true. Ernst Mayr, a noted evolutionist and Professor Emeritus at Harvard, said, "No educated person any longer questions the validity of the so-called

---

[1] Hans Christian Andersen, "The Emperor's New Suit," in *The Complete Hans Christian Andersen Fairy Tales*, ed. Lily Owens (New York: Portland House, 1997), pp. 438-443.

theory of evolution, which we now know to be a simple fact."[2] However, we know the tide is turning. Besides the 50 percent of Americans that reject evolution, there are thousands of scientists that have abandoned evolution and are calling out that the emperor is naked. Dr. Russell Humphreys, a physicist at the Sandia National Laboratories in New Mexico has estimated that there are as many as 10,000 practicing scientists in the United States that are young earth creationists.[3] Are these the 10,000 brave children willing to proclaim the emperor is naked?

In the fall of 1996, we visited a well-respected geneticist with a Harvard Ph.D. who had taught and researched for over 25 years. We were writing our first book at the time and were intrigued by the absolute lack of any substantial foundational work that supported evolution. You will hear much dogmatic confident assertions about the plausibility of evolution and how it is almost a fact. However, it is extremely doubtful if you can find any real substantial evidence. The reason you won't find any is because none exists — kind of like the emperor's new clothes.

"We are writing a treatise and trying to find some evidence supporting evolution, and can't find any," we said anticipating getting thrown out of the professor's office.

"Neither can I," was his reply as he peered over his glasses.

What did this mean? Isn't this man an icon in the world of research and academia? He has spent over 25 years entrenched in the midst of one of the great bastions of evolutionary thought — the university. And now, those words "Neither can I [find evidence for evolution]." This obviously didn't affect our thinking at all regarding the theory of model of evolution or creation, having already resolved that years before without a doubt. However, what did change was our understanding of how people were dealing with an unsupported

---

[2] Ernst Mayr, "Darwin's Influence on Modern Thought," *Scientific American*, July 2000, p. 83.
[3] C. McCulley, "By the Book," *Crystal City, etc,* Fall 1996, pp. 31-32.

theory. You see, just like the Hans Christian Andersen story, even after the emperor realized that he was in fact naked, he continued on with the procession not willing to admit the folly of the episode.

Fortunately, there seems to be an increasingly turning tide regarding scientists willing to express their realizations that evolution is not valid. Michael Behe, professor of biochemistry at Lehigh University, wrote a book entitled *Darwin's Black Box* in 1997. Besides his intellectual attack and criticism of evolution, he quotes some other distinguished scientists that are questioning evolution as well. One of the best quotes in the book is from Lynn Margulis (Distinguished University professor of Biology at the University of Massachusetts). She is widely known for her theory on the origin of mitochondria (a so-called powerhouse inside our cells). She calls neo-Darwinism "a minor twentieth-century religious sect." Behe reports that at one of her talks, she actually challenged the audience to come up with an unambiguous example of the formation of a new species by the accumulation of mutations. A challenge, that as of the writing of his book has gone unmet.[4] Behe also quotes Jerry Coyne (ironically from the Department of Ecology and Evolution at the University of Chicago) as saying "We conclude — unexpectedly— that there is little evidence for the neo-Darwinian view: its theoretical foundations and the experimental evidence supporting it are weak."[5] Behe himself provides one of the most poignant quotes when he said "The result of these cumulative efforts to investigate the cell—to investigate life at the molecular level—is a loud, clear, piercing cry of design."[6]

Conclusion

This is only the tip of the iceberg. Books like *Evolution: A Theory in Crisis* by Dr. Michael Denton; *Evolution: Challenge of the Fossil*

---

[4] Michael J. Behe, *Darwin's Black Box* (New York: Touchstone, 1996), p. 26.
[5] Behe, p. 29.
[6] Behe, p. 232.

Record, by Duane Gish; and many others are objective criticisms of evolution. What is the reaction to these criticisms — these cries that the emperor is naked? Why will some react as the emperor and his court did by continuing on with the procession — continuing on with believing the evolution theory? Is it fear of ridicule — they might be called stupid or unworthy subjects? Is it because by rejecting evolution means they will be faced with accepting and acknowledging a Creator? That being the case, there is a Divine order and we are all part of it. It adds a dimension of being accountable to our Creator for our actions. If this is true, then we must be prepared to deal with people with sensitivity. It is more than an academic debate. It is really Truth versus a terrible lie.

Robert and Elizabeth Ridlon

Part Two: The Models

CHAPTER FIVE

Evolution Foundations

Basically there are only two main questions that evolution or creation science seeks to answer: The first is how did life originate? The second is why do we see diversity or different types of life forms? In this chapter we will examine the foundational concepts and beginnings of the evolution theory and try to understand what makes evolution an attractive and seemingly logical alternative to creation. In subsequent chapters, we will further discuss and evaluate the assumptions and premises of evolution considering the underlying science (*e.g.,* biology, geology, paleontology, etc.) in more depth.

Participants in the debate on evolution and creation are numerous and beyond the scope of this book to review. There are a myriad of prominent scientists, philosophers, and theologians (both historically and presently) that have weighed in heavily on one side of the argument or another with great zeal. Unfortunately, authoritative contributions such as these can keep a debate alive, but won't win arguments. Today the issues are debated as strongly as in the time of Darwin. The creation scientists number in the thousands and the creation literature relentlessly produces volumes of literature that contradicts evolution and reports the gaping holes in evolutionary thought. At the same time, the evolution proponents continue to produce "evidence" from discoveries that pronounce victory, having discovered something new proving their theory. However, there are some contributions that represent the key components of each position; therefore, we will include these references.

The Beginning of Evolutionary Thought

Evolution allows for and explains the beginning of the very first living organism — the origin of life; and the beginning of new kinds of organisms — the diversity of life. The evolution model states that life originated spontaneously from non-life, forming the common ancestor of all living things. All living things that have ever lived have that common ancestor and, from that point, evolved over time into the organisms that have populated the earth. This literally means that the daisy, the fly, the dog, the mushroom, the oak tree, the human, and every other living thing, are related by a common primordial ancestor.

We usually attribute the concepts of evolution to Charles Darwin, who published his theories beginning in 1859. However, Darwin didn't have the original idea. Evolutionary thought probably had its earliest recorded origins from classical Greek philosopher scientists, such as Anaximander of Melitus (c.611-547 BC) and Empedocles (c.493-433 BC). Anaximander was a mathematician and astronomer who had a number of ideas about both the origin of life and the origin of the universe itself. He actually theorized that life arose from certain elements as a naturalistic explanation of the origin of life. Empedocles later proposed that humans and animals had evolved from a common primitive form. He suggested that matter was able to reform into various shapes and the resulting organisms, if fit, would be able to survive. These ideas were very much alive and well up to the time of Darwin. Even Charles Darwin's grandfather, Erasmus Darwin, had written philosophically about nature-driven origin and diversity. Also, Darwin himself was quite aware of the scientists and other intellectuals that were publishing work in support of the concept of life descending from a common ancestral primordial form. So what we often like to attribute exclusively to Darwin, can actually be traced back some 2500 years earlier, and had a number of other proponents along the way.

Charles Darwin was born in Shrewsbury, England in 1809, the son of a practicing physician. Charles had an interest in certain aspects of science and attended medical school for two years at Edinburgh. However, he didn't want to be a physician and considered his time there as wasted. His father suggested the clergy and sent Charles off to Cambridge. Darwin's idea was to live off of the expected inheritance from his father, and studying for the clergy at Cambridge was just as distasteful as becoming a physician. Darwin admitted his dislike of the lectures and, except for his few bursts of effort, admitted to doing the minimum required to get through.

> Although as we shall presently see there were some redeeming features in my life at Cambridge, my time was sadly wasted there and worse than wasted. From my passion for shooting and for hunting and when this failed for riding across country I got into a sporting set, including some dissipated, low-minded young men. [7]

Despite partying with friends and participating in shooting sports, Darwin did study enough to pass his final examination and graduated in 1831. During his tenure at Cambridge he met some influential scientific intellectuals. He studied geology as a side interest and upon returning from a geology field trip, he was informed of an opportunity to travel aboard the *H. M. S. Beagle* as a naturalist.

This now famous voyage, from 1831 to 1836, took Darwin around the world: Australia, Africa, South America, and to the now famous archipelago — the Galapagos Islands. The actual purpose of the expedition was one of surveying and taking chronographic measurements. Charles Darwin, as a volunteer naturalist, made his observations, assimilated the material of his mentors, and developed

---

[7] Charles Darwin, *Charles Darwin Thomas Huxley Autobiographies,* ed. Gavin de Beer (Oxford: Oxford University Press, 1983), p. 33.

and popularized the ideas that many others already had. Over twenty years later, Darwin wrote and published his famous book *The Origin of Species*, in 1859.

The Diversity of Life Explained By Darwinian Evolution

During the five year voyage of the Beagle, Darwin was exposed firsthand to many biomes (or regions) which exhibited a biodiversity few scientists had seen before. The voyage took Darwin around much of the continent of South America, from the jungles of Brazil to the deserts of northern Chile, the coral reefs on the southern tip of S. America to the Galapagos Islands off the coast of Equador. He observed a great amount of differences between the related species on the islands. For example, island finches (often refereed to as Darwin's Finches) had beaks that were uniquely adapted to a particular feeding strategy necessary for acquiring food peculiar to that biome. And, it wasn't just the beaks that differed. The birds' size and feathers differed as well. Darwin interpreted these finches to be of 14 different but related species of finches; and none of these species existed on the mainland. This led Darwin to the conclusion that each was a new species that had descended from some original type.

Some time after the voyage, Darwin read *An Essay on the Principle of Population* by Malthus suggesting that human populations could grow in number and at some point exceed their food supply unless some disaster or war came along and reduced the population. Darwin applied this concept to other organisms. He theorized that in nature events such as natural disaster or war occurred which would severely reduce the populations of a species of organism. However, the species that had certain variations would be favored, survive, and then pass those favorable variations on to its offspring.

Darwin was very well acquainted with a concept of selection used with domestic animals. This so-called artificial selection was used to select for desirable characteristics or variations in domestic animal species such as wool quality in sheep. The pressures that would be exerted by natural environmental pressure were similar to the selective breeding or artificial selection of domestic animals and agricultural products by man's manipulation. Darwin suggested that the same selective mechanism was at work in nature — or as Darwin put it, *natural selection.* Darwin was suggesting that natural environmental differences might exert pressures over time that favored the variations possible within a species to become the dominant variation in a particular environmental biome. Furthermore, he was actually suggesting that it was the accumulation of many variations, over time, that eventually led to the brand new species arising which is completely distinct from the original. This was very much in contradiction to the widely held conventional (Biblical) maxim of the fixity or immutability of the species.

Darwin drew these conclusions which defined his theory of evolution. There is variation within a species of organisms (*e.g.,* tall or short). These variations can be passed on from one generation to another (*i.e.,* tall parents may have tall offspring). The environment may favor certain variations within a species (*e.g.,* advantageous to be tall). Darwin's theories provide for: (1) Variation within species (*e.g.,* variation as in 150 varieties of dog, but all the same species); (2) New species arising within families of organisms due to nature favoring accumulated variations over time; (3) Unrelated species are a result of the same persistent natural selection acting on variations over time (*i.e.,* dogs were once frogs).

The Origin of Life Through Evolution

Another principle of evolution is that nonliving material goes to living material; thus the origin of first life through natural processes. This is a basic premise of evolution since the idea of a Creator or intelligent designer (*i.e.,* God) is dismissed and instead, the very first life was generated by some set of naturalistic circumstances. Darwin didn't really deal with this principle in his research and writing. He accepted the concept that life could have arisen from the primordial slime of a newly formed earth, but didn't explore that aspect of evolution. Darwin is quoted in Denton:

> It is often said that all the conditions for the first production of a living organism are now present which could ever have been present. But if (and oh! what a big if!) we could conceive in some warm little pond with all sorts of ammonia and phosphoric salts, light, heat, electricity present that a protein was formed ready to undergo still more complex changes at the present day such matter would be instantly devoured or absorbed which would not have been the case before living creatures were formed.[8]

One of the earliest alternatives to creation as an explanation of how life may have originated, called *spontaneous generation*, was popular until the mid-1800s. Spontaneous generation was largely based on the observation of flies "arising" from meat and rats emerging from piles of rags and garbage. Nobody saw the flies lay the eggs on the meat, only the emerging larvae and flies. This misconception persisted even after an Italian physician named Francesco Redi disproved it beginning with some experiments in 1668. These experiments of Redi demonstrated that flies actually laid

---

[8] Michael Denton, *Evolution: A Theory in Crisis* (Bethesda, Maryland: Adler and Adler, 1986), p. 53.

33

the eggs on uncovered meat. The eggs hatched into the larvae that grew into mature flies. A contemporary of Redi, Anton Van Leeuwenhoek, described microorganisms he saw through the newly discovered microscope. This led many scientists to suggest that these "simple" organisms could be generated spontaneously.

One convincing experiment supporting spontaneous generation was conducted by John Needham in 1745. It involved heating a nutrient broth, thus killing the microbes. However, a short time later, organisms were again seen in the flasks. This held up in favor of spontaneous generation for twenty years, until Lazarro Spallanzani suggested that the microbes may be entering (*i.e.,* contaminating) the broth through the air. He suggested sealing the containers after the heating process. Sure enough, the microbes were not present in the heat-treated and sealed containers. Needham's response was that there was a "vital force" that was eliminated in the heat treatment and wasn't able to return due to seal on the containers. It was also suggested that the sealed containers eliminated the oxygen needed to support life.

In 1858, a Rudolf Virchow proposed his theory of biogenesis, which proposed that life came only from life and that cells only came from other cells. It wasn't until the French scientist Louis Pasteur executed his experiments, that the issue was nearly resolved. Pasteur's experiments were able to account for all the criticisms. He designed an apparatus in which the nutrient broth was heat-treated, thus killing the existing organisms. His apparatus was U-shaped which allowed the air (oxygen) to enter, but trapped the contaminating particles from entering the broth. This proved that only life gave rise to life and should have sent a message that life didn't arise from the primordial soup of the early earth.

Despite the work of Pasteur and others, there was a resurgence of interest in life originating from non-life in the 20th Century. A Russian scientist named Oparin proposed a chemical theory that once again suggested that certain primordial conditions and events could theoretically give rise to life. There has been a great deal of interest in trying to prove that theory since Oparin's suggestion in 1924. The most notable was that of a scientist at the University of Chicago named Stanley Miller, who in the 1950s developed a recipe for the primordial soup and built an apparatus to produce life or components of life. Unfortunately, neither he nor any scientists since, have been able to come up with a successful experiment that produces life in any form.

Evolution Assumptions

Much has been written describing the conditions that gave Darwin confidence in writing of his controversial work. We believe there were two main contributing factors. First, Darwin needed the earth to be very old in order for his theory to be possible (*i.e.,* millions of years). His ideas of evolution of new diverse species due to natural selection would require an enormous amount of time — a very old earth. The Biblical view considered the earth to be quite young (*i.e.,* less than 6,000 years old). Evolutionists generally agree that it takes a long time for evolution to occur. This is the main reason for the claims of the earth being so old (perhaps 4.5 - 5 billion years or so). However, there is much evidence for a young earth.

The Biblical concept of an earth less than 6,000 years old was unacceptable to Darwin since it didn't provide enough time for the sorts of changes to take place that Darwin proposed. In the late 1700s, a Scottish geologist named James Hutton, proposed that the sedimentary deposits seen everywhere were made by the slow gradual steady deposits over time. Hutton, considered the father of modern geology, hypothesized that the geological processes that we see operating today were operating the same way in the past. This hypothesis is known as uniformitarianism and suggests that slow sedimentation is responsible for the thousands of meters of sedimentary rock comprising much of the geology of the earth. This didn't exclude the possibility of episodic catastrophes (*e.g.,* volcanoes, earthquakes, etc.) but explained the deep layering by slow uniform steady processes of silt deposition taking millions of years — which was more than enough time needed by Darwin and others for evolutionary processes.

A geologist named Charles Lyell, helped Darwin here. Darwin took Lyell's newly published (1830) work entitled *Principles of Geology* with him on the voyage. In this volume, Lyell developed Hutton's hypothesis and explained geologic formations in terms of gradual processes that took a very long time and were carried out by natural, non-catastrophic events. This so-called uniformitarianism went against the grain of the Biblical constructs of a young earth. This component of Darwin's view of geologic time turned out to be a tremendous gift to Darwin who had an enduring friendship with Lyell.

> I am proud to remember that the first place, namely St. Jago, in the Cape Verde Archipelago, which I geologised, convinced me of the infinite superiority of Lyell's views over those advocated in any other work known to me.[9]

Darwin himself may shed some light on the second factor contributing to his exposition concerning evolution. In his autobiography, Darwin suggests that he struggled with the idea that unbelievers in Christ would perish. He also said that the Old Testament "was no more to be trusted than the sacred books of the Hindoos [sic], or beliefs of any barbarian."[10] He goes on to say that he felt driven to the conclusion that "The old argument from design in nature...fails, now that the law of natural selection has been discovered."[11] By his own admission, Darwin has rejected the tenets

---

[9] Darwin, *Autobiographies,* p. 59.
[10] Darwin, *Autobiographies,* p. 49.
[11] Darwin, *Autobiographies,* p. 50.

of Biblical Christianity and admits that he was never firmly convicted of the existence of God.

What Darwin was thinking cannot be assessed definitively. Many authors have tried to read Darwin's thoughts as they have unfolded in the pages of his journals, letters, autobiography, and of course his books on evolution. We can see conflict as he struggled with understanding and interpreting what he saw in nature.

Conclusion

Darwin's observations, analysis, and years of thought culminated in the publication of a theory that attacked the creationists' doctrine of the *fixity of the species*. At the time of Darwin, fixity of species was the predominant belief among scientists and intellectuals. Darwin now suggested that over time, nature's pressures could select for variations in species and eventually, through the accumulation of these favorable variations, give rise to an entirely new species. All things were then related back to a primordial first life and gradual slow processes produced each and every species that has ever existed on this planet, including humans.

Darwin's theory on evolution doesn't deal with origin of life directly. However, the premises of his model provided traction for contradiction of the creationists' position of God as Creator of first life and kept a door open for the concept of life originating from non-life under the natural conditions.

The mechanism that produced the variations in species and the mechanism for transmitting those variations to successive generations wasn't known by Darwin. At the same time Darwin was working on his evolution theory (1838-1859), Gregor Mendel was exploring the mechanisms of inheritance (*i.e.,* genetics). Darwin thought that environmental factors (including use and disuse of a particular characteristic) somehow had an influence on the reproductive organs of the parents. This environmental influence was then passed on to the offspring which then exhibited those same variations and would thereby retain variations that could be selected for, as favorable, or selected against as unfavorable. For example, if a man became muscular by chopping wood or some other physical activity, he could pass that strength (as a trait) to his offspring. At the time of the writing of *Origin of Species*, Darwin believed the variations observed within a species were somehow acquired.

> Deviations of structure are in some way due to the nature of the conditions of life, to which the parents and their more remote ancestors have been exposed during several generations...the reproductive system is eminently susceptible to changes in the conditions of life; and to this system being functionally disturbed in the parents, I chiefly attribute the varying or plastic condition of the offspring.[12]

Another problem with Darwin's ideas was the lack of evidence for transitional or intermediate forms. There is no evidence in the fossil record for transitional species. The following long passage from Darwin's *Origin of Species* expresses his concern for this question:

> On this doctrine of the extermination of an infinitude of connecting links, between the living and extinct inhabitants of the world, and at each successive period between the extinct and still older species, why is not every geological formation charged with such links? Why does not every collection of fossil remains afford plain evidence of the gradation and mutation of the forms of life? We meet with no such evidence, and this is the most obvious and forcible of the many objections which may be urged against my

---

[12] Charles Darwin, *The Origin of Species by Means of Natural Selection or the Preservation of Favoured Races in the Struggle for Life*, ed. J. W. Burrow (New York: Penguin Books, 1987), p. 173.

theory. Why, again, do whole groups of allied species appear, though certainly they often falsely appear, to have come in suddenly on the several geological stages? Why do we not find great piles of strata beneath the Silurian system, stored with the remains of the progenitors of the Silurian groups of fossils? For certainly on my theory such strata must somewhere have been deposited at these ancient and utterly unknown epochs in the world's history.[13]

In the next chapter, we will look at a revised view of evolution in light of these unanswered questions.

---

[13] Darwin, *Origin of Species,* p. 438.

CHAPTER SIX

Evolving Evolution

Darwin left three big questions on the table. The first two involved the cause of variation and the mechanism of inheritance. How did variations occur and how were they passed on from parents to offspring? Darwin suggested that these variations were the characteristics upon which natural selection exerted its pressure (either favorably or unfavorably). However, he attributed the variations to a mechanism of acquisition whereby use or disuse of a particular structure caused it to undergo change. He postulated that this would somehow affect the reproductive system of that organism and then pass that trait to the offspring. This wasn't a bad idea, but it was incorrect. In the upcoming discussion, we will see how Mendelian genetics easily answers the question of how traits (or variations) are transferred from parents to successive generations. However, the second question of how variations arise, is still only addressed by speculation.

There was a third major unanswered question in the minds of many of Darwin's contemporaries (as well as Darwin himself). If evolution was a slow gradual process, and a new species was the result of the accumulation of changes, where are the intermediate forms that reflect those small incremental changes? There were no intermediates coexisting with the forms they were displacing. There were no intermediates found in the fossil record. So how can this be explained? Resolution of this has become a difficult quest for the evolutionist.

The aforementioned issues (*i.e.,* cause of variations and lack of transitional forms) have caused evolution theory to evolve as it attempts to explain these perplexing issues.

Variation and Inheritance Explained by Evolution

With the advent and understanding of the principles of inheritance and the so-called Mendelian genetic principles (see Appendix C), the component of Darwinian theory that explained variation in terms of use and disuse was abandoned. Today both creationists and evolutionists know the mechanism of variation is gene-based. Scientists (as well as grade-school children) know that there is variation in the way people (or any organism) look and behave. The DNA in each and every cell of an organism defines that organism in terms of what species it is (dog or cat or tree), but also what it looks like (*e.g.,* short or tall, light or dark, etc.). This DNA is the molecular structure comprised of the genes that code for these defining characteristics. The genes are what get passed from each parent organism (*e.g.,* mother and father) to their offspring thereby passing on their "variations."

Darwin's idea of acquired characteristics was definitely wrong, but we still haven't answered the question of why we see any differences at all (*i.e.,* variations). Why aren't all dogs black Labrador retrievers? We know the variations exist, but how do they come about? The answer given by the neo-Darwinian evolutionists is that the variations are a result of random mutations of the genes that were responsible for the characteristics (*e.g.,* hair color, size, etc.). It is the random mutation that is responsible for the variations in organisms and these variations were either good or bad for the organism's survival depending upon the natural selective conditions of the environment — thus survival of the fittest.

Although, random mutation seems to be a good answer, there are some serious concerns from a practical standpoint. As we more closely examine this and other components of evolution in upcoming chapters, we will see there are many serious flaws in this approach. Mutations themselves do occur; however, they are generally inherently harmful. The mutation rate in cells is very low and to have a combination of meaningful mutations that would give the kinds of selective advantages that equates to evolution is next to impossible. This is another example of a theoretical explanation that has gone out of favor. Although it still persists in textbooks today, the idea of accumulated mutations being the mechanism for evolution has lost support by some of evolution's greatest advocates.

Lack of Intermediate Forms Explained

Today, Darwin's original premises of natural selection, preservation of the favorable variations and rejection of so-called injurious variations, still holds sway with evolutionists. As we saw in the previous section, the neo-Darwinists figured out Mendelian genetics and corrected the mechanism by which the variations are passed on to successive generations. Despite the obvious difficulties with the concept of mutations being the mechanism for change, this is still a widely held view. There is still another perplexing issue that plagues the evolutionist: Where are the intermediates or transitional forms? This problem was one that Darwin himself faced and was hard pressed to explain. If variations were selected for (or against) in the Darwinian sense of the struggle for survival, there should be evidence in the fossil record of the failures, as well as the intermediates between the original species and the resulting new species as would occur over time.

Darwin couldn't definitively answer the questions over a hundred years ago. He was heavily criticized for the lack of an explanation. He considered the imperfections in the geological record, but said "Grave as these several difficulties are, in my judgement they do not overthrow the theory of descent with modification."[14] Since the time

---

[14] Charles Darwin, *The Origin of Species by Means of Natural Selection or the Preservation of Favoured Races in the Struggle for Life*, ed. J. W. Burrow (New York: Penguin Books, 1987), p. 440.

Darwin wrote these poignant words, there has been a relentless attempt at finding the multitude of transitional forms that should be present in the fossil record. There have been a precious few candidates; and these have been nearly discounted. This has required the evolutionist to once again redefine the theory of evolution from neo-Darwinism to something like neo-neo-Darwinism.

Darwin was insistent upon the slowness of the transitional process:

> That natural selection will always act with extreme slowness, I fully admit.[15]

> Slow though the process of selection may be, if feeble man can do much by his powers of artificial selection, I can see no limit to the amount of change, to the beauty and infinite complexity of the coadaptations between all organic beings, one with another and with their physical conditions of life, which may be effected in the long course of time by nature's power of selection.[16]

Therefore the geology, in the Hutton/Lyell sense (old earth), ought to reveal the intermediate forms as they transition from one species to another — but, it doesn't. So where is the evidence?

One early explanation was called the "hopeful monster" by Richard Goldschmidt in 1933. He published a paper where he described the possibility that a multitude of things would happen in a single genetic step or jump that would endow the organism with everything necessary to give it an adaptive advantage in a new environmental niche.[17] The type of genetic leap described by Goldschmidt requires a tremendous (and frankly impossible) amount

---

[15] Darwin, *Origin of Species,* p. 152.
[16] Darwin, *Origin of Species,* p. 153.
[17] Richard Goldschmidt, *The Material Basis of Evolution* (New Haven: Yale University Press, 1940), p. 390.

43

of simultaneous change in an organism. The change would have to account for the huge gaps currently seen in the fossil record. This idea doesn't hold much credibility among modern evolutionists.

Another explanation is suggested by the prominent evolutionary paleontologists Stephen Gould and Niles Eldridge. They published their alternative to gradualism in the early 1970s, proposing a model called *punctuated equilibria*. In this model, evolution takes place in peripherally isolated groups that live apart from the parental stock. A new species arises within this isolated area and rapidly forms a population (*i.e.,* peripheral population). As the presumed advantaged species, it then spreads out over a large geographic area leaving behind the evidence of the transition only in the small geographic region. Thus, it would be hard to link the new species with the old original one from which it was derived. Gould wrote:

> A new species can arise when a small segment of the ancestral population is isolated at the periphery of the ancestral range...What should the fossil record include if most evolution occurs by speciation in peripheral isolates? Species should be static through their range because our fossils are the remains of large central populations. In any local area inhabited by ancestors, a descendant species should appear suddenly by migration from the peripheral region in which it evolved. In the peripheral region itself, we might find direct evidence of speciation, but such good fortune would be rare indeed because the event occurs so rapidly in such a small population. Thus, the fossil record is a faithful rendering of what evolutionary theory predicts, not a pitiful vestige of a once bountiful tale.[18]

This makes some sense; however, there has yet to be a paleontological discovery that substantiates this approach.

---

[18] Stephen J. Gould, *The Panda's Thumb* (New York: W.W. Norton and Co., 1980), pp. 183-184.

Furthermore, Michael Denton, a medical doctor and scientist, points out that that the large gaps that would exist between a terrestrial mammal and a whale, for example, would have had required long lineages and collateral lines of transitional forms. He finds the idea incredible that the thousands or even millions of necessary transitional forms as intermediates would have been restricted in geography and population size.[19]

Origin of Life Explained by Modern Evolutionary Thought

Scientists wanted to know how organic compounds and DNA were formed in order to advance hypotheses on how the first cell evolved. A. I. Oparin, a Russian scientist in 1924, provided evolutionists with an hypothesis called *chemical evolution* that explained how the first life was formed from organic chemicals. The theory stated that inorganic molecules in gaseous form would have reacted in the earth's primitive atmosphere to form organic molecules such as amino acids. The organic molecules then would have reacted with each other to form polymers. For example, amino acids would be formed first, then react with one another to form a polymer of amino acids, which is called a protein. Likewise, simple sugar molecules would form a polymer known as a carbohydrate. Also, nucleotides would have reacted with each other to form the nucleic acid polymer - (DNA or RNA). From this point, the polymers would have self-assembled to form a complete living reproducing cell.

In the early 1950s, Stanley Miller, a graduate student of chemistry at the University of Chicago, decided to test Oparin's hypothesis. He set up a glass apparatus where gases carried by steam are sparked by electricity to simulate the early earth atmosphere. The gases used,

---

[19] Michael Denton, *Evolution: A Theory in Crisis* (Bethesda, Maryland: Adler and Adler, 1986), pp. 193-194.

which are simple raw materials thought to occur in earth's primitive atmosphere, are hydrogen, methane, ammonia, and water vapor. After a few days there was a brown liquid that contained many organic compounds including amino acids, urea, and lactic acid. Other experiments found that nucleotides could also be formed in the lab from inorganic compounds. Scientists have also been able to form the polymers (protein, nucleic acids, and carbohydrates). Therefore, it appeared that Oparin's hypothesis could have occurred on the earth. However the DNA formed does not contain any genes that are used in living cells.

Scientists still cannot explain how polymers might self-assemble to form a living cell. Henry Morris developed a statistical argument that supports the improbability of life originating by chance, even with generosity to the evolutionist (an old universe, an extremely simple organism, and an extremely rapid rate of particle assemblage). Here is the simple illustration:[20]

Spontaneous Origin of Life Probability

Generous Assumptions

- The universe is 30 billion years old (10^{18} seconds).
- The first organism has only 100 integrated parts.
- There are 10^{80} available particles that can be used (based on the estimated number of electrons in the universe). It isn't the electrons that would be linking together. The reference to electrons is only to be generous in estimating the number of available parts for the 100-part organism. Each linking would yield 10^{78} combined forms of 100 particles each.

---

[20] Henry M. Morris, *Scientific Creationism* (El Cajon: Master Books, 1985), pp. 60-61.

- Linking and unlinking of particles occurs every billionth of a second (1 event per 10^{-9} seconds).
- There is only one correct way the 100 parts can link together. Law of Probability says that the chance of getting the 100 particles linked together correctly is 1 in 100 factorial (1 x 2 x 3 x 4 x ...100) or 1 in 10^{158}.
- Unsuccessful linking would result in unlinking and retrial every 10^{-9} seconds (billionth of a second) until success is achieved.

Calculation

10^{78} combinations x 10^9 combinations per second x 10^{18} seconds = 10^{105} combinations. The chance that one of these 10^{105} is correct is 1 in $10^{158}/10^{105}$ or 1 in 10^{53}. 10^{53} is a one with 53 0s. So, for the non-mathematician, if something has one chance in 10^{53} of occurring, that is really a zero chance.

Morris's calculations, and the contributions of several other studies, conclude that the chance of life assembling itself by random chance events is zero. In fact, when the extremely generous assumptions are defined more realistically, the argument for random assemblage of particles become even more compelling as impossible.

Conclusion

Modifications to Darwin's original ideas have given rise to the concepts of genetic mutations as the mechanism for variation. The creationist's view is that the genetic potential for the variations already exists in the repertoire of genetic material in the cells of living things. Variations are expressions of this repertoire (*e.g.,* blue eyes or brown, tall or short).

The lack of transitional forms was especially troubling to Darwin, as well as today's evolutionists. Gould has called this lack of fossil evidence "the trade secret of paleontology."[21] The punctuated equilibrium concept did little to satisfy the need for finding missing links, but nonetheless provided an excuse (correct or not) for the lack of one of the biggest pieces of evidence needed by evolution.

So far, no life has been created in the lab. Scientists today believe the Law of Biogenesis (*i.e.,* life comes from life), but with one exception. They think that in the very beginning, new life (a cell) arose from organic polymers. If it occurred many years ago, then it should still be occurring at present. However, we see no evidence of this anywhere on earth. We do not see new cells arising from the oceans or ponds nor are we able to create it artificially in the lab.

In the next chapter, we will look at the creation view of the same questions of life's origin and diversity. The creationist, on the other hand, believes that the DNA (genes) within a population of the same species, has variation built in (God-given). This variability is a repertoire of genetic variability that enhances an organism's potential for survival under the pressures of natural selection.

---

[21] Gould, pp. 181.

CHAPTER SEVEN

Foundations of Biblical Creation

As we said in the previous chapters on evolution, there are two questions that are addressed by both creationists and evolutionists: The first is how did life originate? The second is why do we see diversity or different types of life forms? The answer to these two questions establish the main points of contention between creation and evolution. If there is one area where the proponents of evolution and the proponents of creation agree, it's that these points are at the center of the argument. In this chapter we will look at the Biblical foundations of the creationists view. In the next chapter, we will consider these foundations within a discipline called creation science.

The basic premise of the Biblical creation model is found in Genesis 1:1 which states that "In the beginning God Created the *heavens* and the *earth*." The Apostle Paul restates the belief in his letter to the Colossians:

> For by him all things were created: things in heaven and on earth, visible and invisible, whether thrones or powers or rulers or authorities; all things were created by him and for him. (Colossians. 1:16)

There are two main principles that can be derived from the creation model. First, life was created through an intentional act of God — not by chance. Second, the diversity (or differences) among organisms is attributable to God's original creation. Although there is variability within species (*e.g.,* poodles and collies), this doesn't give rise to new species (*e.g.,* frogs didn't become dogs).

The Bible Outlines Creation

Understanding the creation model requires us to first examine the scriptures that describe the process of creation. This will serve as the foundation for the creation model.

The First Day of Creation: Heavens, Earth, and Light

1 In the beginning God created the heavens and the earth. 2 Now the earth was formless and empty, darkness was over the surface of the deep, and the Spirit of God was hovering over the waters. 3 And God said, "Let there be light," and there was light. 4 God saw that the light was good, and he separated the light from the darkness. 5 God called the light "day," and the darkness he called "night." And there was evening, and there was morning —the first day. (Genesis 1: 1-5)

From the first verse in the Bible, we learn that God created the heavens and the earth. This first verse is sometimes considered an overview of creation. The word *heavens* can be interpreted in a number of ways. In this context, heavens probably refers to the whole universe that exists apart from the earth. In this first verse we see the earth was created as well. This is the world we dwell in as opposed to the heavens. In the next verses, we see the third object of creation is light. An interesting note here is that the sun, moon, and stars have not yet been created, yet there is light. One can only speculate, but it seems that light existed before the sun and moon were "assigned" the job of providing light (Genesis 1:14-19).

God formed the creation from His command. This creation principle from Genesis is presented by the writer of Hebrews:

By faith we understand that the universe was formed at God's command, so that what is seen was not made out of what was visible (Hebrews 11: 3).

This means that everything was created *ex nihilo* (from nothing) and God was the Designer and the Creator.

Second Day of Creation: Sky and Vapor Canopy

6 And God said, "Let there be an expanse between the waters to separate water from water." 7 So God made the expanse and separated the water under the expanse from the water above it. And it was so. 8 God called the expanse "sky." And there was evening, and there was morning—the second day. (Genesis 1: 6-8)

Verses 6 and 7 both refer to an expanse which separated two waters—the waters above and the waters below. These waters were part of the formless void matter created in verses 1 - 5. The waters below the expanse seem easily understood as the earth's surface water, but what about the waters which were above the expanse? This is an important concept which is referred to as the vapor canopy.[22] This might be thought of as water vapor or cloud layer above the earth. It may have served as a shield from direct sunlight and the sun's harmful ray and formed a virtual greenhouse effect on the earth resulting in warm uniform temperatures. There would be no rain and no polar cold conditions. The Bible and science both provide some solid support for this concept, which we will examine further in the next chapter.

---

[22] John C. Whitcomb and Henry M. Morris, *The Genesis Flood* (Philadelphia: The Presbyterian and Reformed Publishing Company, 1961), pp. 121, 254-257.

Third Day of Creation: Land, Seas, and Vegetation

9 And God said, "Let the water under the sky be gathered to one place, and let dry ground appear." And it was so. 10 God called the dry ground "land," and the gathered waters he called "seas." And God saw that it was good. 11 Then God said, "Let the land produce vegetation: seed-bearing plants and trees on the land that bear fruit with seed in it, according to their various kinds." And it was so. 12 The land produced vegetation: plants bearing seed according to their kinds and trees bearing fruit with seed in it according to their kinds. And God saw that it was good. 13 And there was evening, and there was morning — the third day. (Genesis 1: 9-13)

We see that on the third day the land and seas were gathered into one place. They were previously created on the first day as the formless earth. This implies that there was only one landmass, not the separate seven continents and islands we have today. There is evidence of one landmass in the past that geologists call *pangaea*.[23]

The creation of vegetation, the plants and trees, is recorded in verse 12. This is the first living thing created. Notice the phrases "according to their kinds." The plants and trees made seeds and those seeds would grow another tree just like its parent. The concept of fixity of species is derived from this phrase and its context.

Fourth Day of Creation: Sun, Moon, and Stars

14 And God said, "Let there be lights in the expanse of the sky to separate the day from the night, and let them serve as signs to mark seasons and days and years, 15 and let them be lights in the expanse of the sky to give light on the earth." And it was so. 16 God made

---

[23] Charles C. Plummer and David McGeary, *Physical Geology* (Dubuque: W. C. Brown Company, 1982), p. 388.

two great lights, the greater light to govern the day and the lesser light to govern the night. He also made the stars. 17 God set them in the expanse of the sky to give light on the earth, 18 to govern the day and the night, and to separate light from darkness. And God saw that it was good. 19 And there was evening, and there was morning - the fourth day. (Genesis 1: 14-19)

On the fourth day God created the sun, moon and stars. God first created the sun and the moon to give us light during the day and the night. Their other purposes were to create a daytime and a nighttime and to mark off the seasons and the years. God actually fixed a calendar for us. Our calendar is based on the revolution of the earth around the sun. God also created the stars at this time.

Fifth Day of Creation: Marine Life and Birds

20 And God said, "Let the water teem with living creatures, and let birds fly above the earth across the expanse of the sky." 21 So God created the great creatures of the sea and every living and moving thing with which the water teems, according to their kinds, and every winged bird according to its kind. And God saw that it was good. 22 God blessed them and said, "Be fruitful and increase in number and fill the water in the seas, and let the birds increase on the earth." 23 And there was evening, and there was morning- the fifth day. (Genesis 1: 20-23)

We once again see this phrase "according to their kinds" with reference to the created organisms, this time applied to the marine life and the birds.

Sixth Day of Creation: Various Land Animals, Creatures that Move on the Ground, and Man

24 And God said," Let the land produce living creatures according to their kinds: livestock, creatures that move along the ground, and wild animals, each according to its kind." And it was so. 25 God made the wild animals according to their kinds, the livestock according to their kinds, and all the creatures that move along the ground according to their kinds. And God saw that it was good. 26 Then God said," Let us make man in our image, in our likeness, and let them rule over the fish of the sea and the birds of the air, over the livestock, over all the earth, and over all the creatures that move along the ground." 27 So God created man in his own image, in the image of God he created him; male and female he created them. 28 God blessed them and said to them, "Be fruitful and increase in number; fill the earth and subdue it. Rule over the fish of the sea and the birds of the air and over every living creature that lives on the ground." 29 Then God said, "I give you every seed-bearing plant on the face of the whole earth and every tree that has fruit with seed in it. They will be yours for food. 30 And to all the beasts of the earth and all the birds of the air and all the creatures that move on the ground-everything that has the breath of life in it- I give every green plant for food." And it was so. 31God saw all that he had made, and it was very good. And there was evening, and there was morning-the sixth day. (Genesis 1: 24-31)

During the sixth day of creation, God created the land animals, which included both wild animals and livestock. The creatures that moved on the ground were also created at this time. This must have included the reptiles, as well as insects and spiders. God also created His centerpiece on day six — man. God made man, in His image, and gave man dominion over all the rest of the creation. It is clear that God made both the male and the female human on this sixth day.

In the descriptions for the events for each of the three days where living things were created (Days three, five, and six), we see the phrase repeated "according to their kinds." In the 1700s, a man named Carl von Linne was the first person who interpreted "kinds" to mean species in biology. He was born in Rashult, Sweden on May 23, 1707. Linne was the son of a pastor and had great respect for the Bible. He became a physician and studied botany. His goal was to name and organize into a nomenclature system all the original Genesis "kinds" in the Bible. He thought the species (kinds) could have variation, but that the species were established and fixed. In other words he did not believe that the species could evolve into another different species. He is considered the father of biological taxonomy called the Linnean system which is still the standard classification system.

In Genesis Chapter 2, we see the details of the creation of man and woman. It begins with a short recapitulation and is followed by an explanation of the events surrounding God's creation of man and woman.

This was apparently a perfect world and was pronounced "good" by God. There was no bloodshed required to get food for either man or beast, since God provided plants as the food source for all. Unfortunately, things did change and the perfect world became imperfect as we will see.

The Fall and the Flood

The Fall and Subsequent Corruptness of Mankind
(Genesis Chapter 3 -6))

17 To Adam he said, "Because you listened to your wife and ate from the tree about which I commanded you, `You must not eat of

it,' "Cursed is the ground because of you; through painful toil you will eat of it all the days of your life. 18 It will produce thorns and thistles for you, and you will eat the plants of the field. 19 By the sweat of your brow you will eat your food until you return to the ground, since from it you were taken; for dust you are and to dust you will return." (Genesis 3: 17-19)

Man rebelled in an act of disobedience. This was an action that had far reaching consequences. From that point on we see changes in the creation. Man's diet is now restricted to the plants of the field which man must work very hard to grow and harvest. This is different from the beginning when God provided the fruit-bearing trees and seed-bearing plants.

God said his creation was good many times in Genesis Chapter 1. Good means not evil, no problems, and perfect. However, we see that due to the fall of man, the creation, along with Adam and Eve, are not perfect any more. In Romans 8:19-22, the Apostle Paul speaks of the creation groaning and undergoing decay.

Things actually became worse after man's initial rebellion. Adam had a murderous son and subsequent generations left the world a wicked place. Only Noah and his family were found to be righteous.

5 The LORD saw how great man's wickedness on the earth had become, and that every inclination of the thoughts of his heart was only evil all the time. 6 The LORD was grieved that he had made man on the earth, and his heart was filled with pain. 7 So the LORD said, "I will wipe mankind, whom I have created, from the face of the earth —men and animals, and creatures that move along the ground, and birds of the air —for I am grieved that I have made them." 8 But Noah found favor in the eyes of the LORD. (Genesis 6: 5-8)

11 Now the earth was corrupt in God's sight and was full of violence. 12 God saw how corrupt the earth had become, for all the people on earth had corrupted their ways. 13 So God said to Noah, "I am going to put an end to all people, for the earth is filled with violence because of them. I am surely going to destroy both them and the earth. (Genesis 6: 11-13)

At this point, God gave Noah the task of preparing an ark which would be used to preserve Noah and his family, along with representatives of all the living creatures. Everything else would be destroyed. God gave Noah the exact specifications for building the ark and also gave Noah exact instructions on the animals that would be brought into the ark.

14 So make yourself an ark of cypress wood; make rooms in it and coat it with pitch inside and out. 15 This is how you are to build it: The ark is to be 450 feet long, 75 feet wide and 45 feet high. 16 Make a roof for it and finish the ark to within 18 inches of the top. Put a door in the side of the ark and make lower, middle and upper decks. 17 I am going to bring floodwaters on the earth to destroy all life under the heavens, every creature that has the breath of life in it. Everything on earth will perish. 18 But I will establish my covenant with you, and you will enter the ark —you and your sons and your wife and your sons' wives with you. 19 You are to bring into the ark two of all living creatures, male and female, to keep them alive with you. 20 Two of every kind of bird, of every kind of animal and of every kind of creature that moves along the ground will come to you to be kept alive. 21 You are to take every kind of food that is to be eaten and store it away as food for you and for them." 22 Noah did everything just as God commanded him. (Genesis 6: 13-22)

Noah was to build an ark 450 feet long, 75 feet wide and 45 feet high. He was to bring animals of every kind (male and female), plus food for Noah's family and all the animals as well.

> The LORD then said to Noah, "Go into the ark, you and your whole family, because I have found you righteous in this generation. 2 Take with you seven of every kind of clean animal, a male and its mate, and two of every kind of unclean animal, a male and its mate, 3 and also seven of every kind of bird, male and female, to keep their various kinds alive throughout the earth. 4 Seven days from now I will send rain on the earth for forty days and forty nights, and I will wipe from the face of the earth every living creature I have made." (Genesis. 7:1)

Noah was to take seven clean animals and of the unclean animals, he was to take two of each kind. The reason for this is unclear here, but some have suggested that some of the clean animals would be used for food and/or possibly for sacrifice. This would allow for the preservation of the species for each of these.

The Flood (Genesis Chapter 7)

God gave very specific instructions concerning how the events would transpire before and during the Flood.

> 11 In the six hundredth year of Noah's life, on the seventeenth day of the second month —on that day all the springs of the great deep burst forth, and the floodgates of the heavens were opened. 12 And rain fell on the earth forty days and forty nights.

> 20 The waters rose and covered the mountains to a depth of more than twenty feet. 21 Every living thing that moved on the earth

perished —birds, livestock, wild animals, all the creatures that swarm over the earth, and all mankind. 22 Everything on dry land that had the breath of life in its nostrils died. 23 Every living thing on the face of the earth was wiped out; men and animals and the creatures that move along the ground and the birds of the air were wiped from the earth. Only Noah was left, and those with him in the ark. 24 The waters flooded the earth for a hundred and fifty days. (Genesis 7: 11-12, 20-24)

The devastation of the Flood effectively eliminated all the people, animals and other creatures. The rain came down for 40 days and nights, but even after 150 days, the flood waters remained. It would be another 150 days before the flood waters were gone. It actually took over a year from the beginning of the rain to the end of the Flood.

Gen. 8:1 But God remembered Noah and all the wild animals and the livestock that were with him in the ark, and he sent a wind over the earth, and the waters receded. 2 Now the springs of the deep and the floodgates of the heavens had been closed, and the rain had stopped falling from the sky. 3 The water receded steadily from the earth. At the end of the hundred and fifty days the water had gone down, 4 and on the seventeenth day of the seventh month the ark came to rest on the mountains of Ararat. 5 The waters continued to recede until the tenth month, and on the first day of the tenth month the tops of the mountains became visible. (Genesis 8: 1- 5)

14 By the twenty-seventh day of the second month the earth was completely dry. 15 Then God said to Noah, 16 "Come out of the ark, you and your wife and your sons and their wives. 17 Bring out every kind of living creature that is with you —the birds, the animals, and all the creatures that move along the ground —so they can multiply on the earth and be fruitful and increase in number

upon it." 18 So Noah came out, together with his sons and his wife and his sons' wives. 19 All the animals and all the creatures that move along the ground and all the birds —everything that moves on the earth —came out of the ark, one kind after another. (Genesis 8: 14- 19)

Conclusion

From the time of Adam and Eve's rebellion, the earth became populated and things occurred which we can't explain with certainty. However, we know that things weren't what God wanted regarding the creation. Therefore, God judged man by natural forces and with a worldwide flood because of man's wickedness. We do not know the exact geological events of the flood. However, there are many models, from a creation science viewpoint. These models are part of what is collectively referred to as the flood model or catastrophism. The geology of the earth today, (*e.g.,* the Grand Canyon, lakes, and mountains) can be explained by the catastrophic event of the flood described in Genesis.

The story of creation as recorded in the Bible is complete in its outline of the events and comprehensive in its coverage of the results. It forms the foundation of what subscribers to Biblical creation believe. It really provides a boundary within which a scientist can operate in seeking to better understand the world and its inhabitants. It doesn't eliminate scientific endeavor. It serves as a defining model that guides anyone that's interested in the truth about origin and diversity of life. The Biblical creation model provides the working space for exploration and discovery.

Science is the quest for knowledge and ultimately truth about our world and our universe. Scientists and scholars employ methods and tools that enable them to see the underlying events and processes that combine to answer the riddles of life itself. Science and its various disciplines (*e.g.,* physics, chemistry, biology, astronomy, geology, etc.) truly endeavor to uncover the truth and should be compatible with the creation model. Observation, analysis, and conclusion, the core of a scientific approach, should apply to the Biblical creation model. This is the subject we will examine in the next chapter.

CHAPTER EIGHT

Creation Science

In watching a recent PBS series on evolution, it was intriguing to see and hear an unfortunate misrepresentation of the creationist viewpoint. Interestingly, it wasn't the evolutionists that were the perpetrators of the misrepresentation; it was the creationists. Even during Darwin's time, some creationists (who were scholarly respected scientists) would dismiss many questions about unknown mechanisms with terse comments such as "That's the way God did it — and who are we to question God." As biologists, we know from our own study of creation science, there is a scientific approach to answering the basic questions on origin and diversity. Creation science doesn't abandon the principles of science and it doesn't throw out any of the methods of analysis. What creation science does is frame the possibilities within a literal Biblical model. There isn't anything unscientific about that. Unfortunately, the creationists that were featured on PBS didn't represent that (or at least what was on the program didn't reflect that concept). Possibly, much of what occurs in the creation/evolution debate is a matter of miscommunication and misunderstanding.

As we saw in the last chapter, the story of creation as recorded in the Bible is both complete in its outline of the events and comprehensive in its coverage of the results. However, belief in a creation model and an intelligent designer, doesn't eliminate scientific endeavor. With creation science, there is an added dimension of faith in supernatural activities: however, the methods of science — observation and analysis are still applicable to the Biblical creation model.

Both the creation model and the evolution model are comprised of numerous submodels, complex arguments, and a fairly large array of points of contention. Collectively, numerous authors have had a great deal of success in developing the case for the creation model over the evolution model or at least raising serious concerns about the validity of evolution. (Notably, Behe's book *Darwin's Black Box*, Denton's *Evolution: A Theory in Crisis*, Gish's *Evolution: Challenge of the Fossil Record*, Morris and Parker's *Scientific Creation*, and the seminal work of Whitcomb and Morris *The Genesis Flood*.) Still, there are two main arguments that define both the creation model and the evolution model. These are the explanation of origin of life and the explanation for the diversity of life. In this chapter we will see how scientific creationism approaches these arguments and provides scientific explanations within the boundaries of the Bible.

Origin of Life Explained by Creation Science

In the six days of creation, which we just examined in the previous chapter, we saw that God created all living things. The Bible didn't say that God created some things and then let nature take its course. It clearly says that each was created according to its kind. Even man was created by God. This clearly defines the origin of life. There is no evidence to the contrary. The oldest civilizations excavated reveal man in the midst of all the creation and in the present modern form.

Diversity of Life Explained by Creation Science

Species and Biblical Kinds

There is little argument that there are observable differences among living organisms. As we briefly discussed in the previous chapter, Carl von Linne devised a classification scheme for arranging living things. This classification system is still widely used today and is discussed in Appendix C of this book. Carl von Linne started by placing organisms in some very broad high-level categories such as animals, plants, etc. From there, each of those categories was subdivided into lower categories that allow the grouping by like characteristics. For example, within the large category of animals, there are some with vertebral columns and some without. Vertebrates can be mammals or fish or reptiles or birds, etc. Within mammals there are dogs and cats and bears and squirrels, etc.

This Linnean taxonomic system was a clever and helpful way to classify things to help with scientific study of organisms. When you get at or near the bottom levels of this scheme, you begin to deal with the absolutely unique classification of species. Carl von Linne was convinced that Biblical kinds meant an organism that was distinct from other organisms. This distinction was based on an organism's ability to reproduce "according to their kinds" as the Bible says. This concept was not only a Biblical idea, it is still the standard for differentiating one species of organism from another. The test for species isn't appearance. Two organisms may look alike and not be the same species (*e.g.,* rats and squirrels). Conversely, organisms may not look alike and, in fact, be of the same species (Chihuahua dogs and St. Bernard dogs). The Biblical model holds. The test is whether or not the organisms are capable of producing viable reproducing offspring — or in other words "according to their kind."

This is a very important point with regard to the Biblical creation model. From time to time there have been odd claims that there are violations of this concept of reproductive isolation. For example, if an organism was classified as *Panthera uncia* (a snow leopard) and it encountered a *Panthera pardis* (an East African leopard) and we witnessed a mating that resulted in viable reproducing offspring, could we then suggest that the Biblical model had been violated? No! We would admit that some well meaning zoologist thought they were different species, but they were in fact the same species. So there we have the concept of species or Biblical kinds. We can count on that.

Variation Within Species

One of the most notable examples of variation within a species is the common pet dog. There are some 150 varieties of dogs that mostly look different from one another, but every dog is a member of the same species — *Canis familiaris*. This means that if the dog is a Chihuahua, it is of the species *Canis familiaris*. If your pet dog is a Dalmatian it is a *Canis familiaris*. If it is a German Shepherd — *Canis familiaris*. Get the idea? Well, if that is the case, then all dogs should be reproductively compatible, right? That is correct and they are reproductively compatible — although the reproductive act is sometimes a complicated endeavor. They are all of the same species (*i.e.,* same Biblical kind) and reproductively capable of producing viable reproducing offspring.

Besides the dog, we see similar, but often less dramatic variation among other organisms of the same species — even in humans. Some of the kinds of variations we see in humans includes height, hair color, eye color, skin tone and color, and many others. People from different lands can even look different in other distinguishing ways. However, we are all the same species (kind) which is *Homo sapiens*.

If you remember the discussions from Chapter 2, this type of variation within a species was very interesting to Darwin. He believed that somehow nature favored some variations and not others and that's how one variety of organism may have dominated a geographic region. He called the pressure that "decided" which variations were favorable *natural selection*. He acknowledged that this idea was based on so-called artificial selection (at the hands of man) when selective breeding or agricultural hybridization was used to optimize a particular trait of variation. The creationist should have no problem with accepting natural selection as it acts on variations. Darwin was merely describing the way a population of organisms can survive under changing conditions.

As genetics students at Indiana University, we studied several examples involving the application of artificial selective pressures on various organisms and the resulting shift in frequency of a particular trait. Some common examples include hybridization of dairy cattle to select for better milk production; breeding of short-legged sheep in England to select for ones that couldn't jump fences; and of course selecting for hens that laid larger (or more eggs). There are limits on how much selective pressure organisms will take. At some point, it will not be possible to go any further. This represents the limits of the genetic material in the population.The point of all this discussion is to acknowledge that natural selection is a reasonable concept that can influence which variety is dominant in a population. This may be thought of as a God given repertoire of genetic potential which enhances the survival opportunity of a population of organisms. What doesn't happen in the Biblical creation model is one species becoming another species.

The Source of Variation

Now to the second component of our discussion on variation — the source or cause of variation. As we have already seen, there can be exhibited a vast, yet finite, amount of variation within a species. Within an individual member of a population, there is a set of traits that, taken together, make that individual unique. For humans, traits like eye color, finger prints, blood type, hair color, etc., are all under the control of the genes in the cells in the body. The traits in an individual organism are essentially fixed and don't change. All living organisms are comprised of genes that have this influence on the exhibited traits for that organism.

On the other hand, a population of organisms will have a spectrum of variation in traits. For example, in a population of gray squirrels, the tail length may vary from 210mm to 240mm with the majority (predominant) variant at 220mm. If suddenly there was a predator that could only catch them by the tail, there could be an adaptive advantage for the short tail squirrels (*i.e.,* ones with tail closer to the 210mm length), and they would soon be the only successfully reproducing variety.

Darwin and others thought that the variation in traits was brought on by external factors applied to the organisms parents which then caused the offspring to be born with some variation. We know this was wrong. The neo-Darwinian view is that genetic mutations occurred, and are occurring, which present alternatives to the various traits and so-called variants can be produced. The Biblical creationist view is that variation is a result of preplanning. If the universe and its inhabitants are created by God (an intelligent designer), it makes sense to build in genetic potential for adaptation.

Mutations do occur in organisms. Unfortunately, these mutations are nearly always fatal or present severe complications that inhibit an organism's life. Genetic mutation should be seen for what it is — an unfortunate consequence of a fallen world exposed to the decay of the universe. It is not a mechanism for advancing organisms up an evolutionary ladder to higher more sophisticated beings.

The Vapor Canopy and the Flood

In the last chapter, we introduced a term and concept called the *vapor canopy*. The so-called vapor canopy and the world-wide flood together form an integral part of the creation model. One reason is that it is the collapse of this vapor canopy that contributed to the vast consuming flooding of the earth thereby accomplishing God's judgement. The lack of the vapor canopy, post-flood, is also an important factor in the way the earth and its environment changed.

Vapor Canopy

As a reminder, here are the scriptures that introduced the vapor canopy:

6 And God said, "Let there be an expanse between the waters to separate water from water." 7 So God made the expanse and separated the water under the expanse from the water above it. And it was so. 8 God called the expanse "sky." And there was evening, and there was morning—the second day. (Genesis 1: 6-8)

The picture here is of an expanse, which is the same sky we have today. However, this expanse or sky is between the surface of the earth and a vast translucent envelop of water vapor. Figure 8-1 is a drawing representing the concept of the so-called vapor canopy.

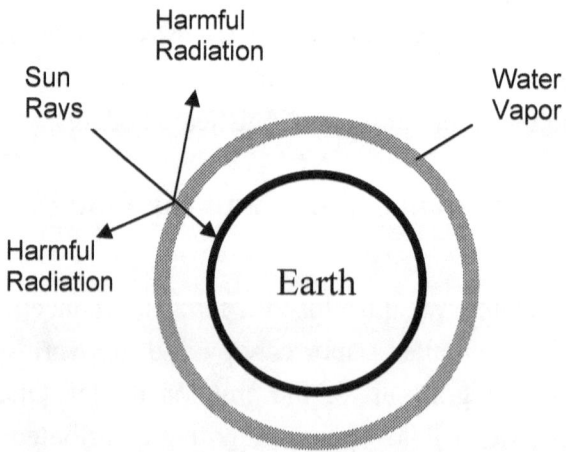

Figure 8-1. Vapor Canopy Concept

This vapor canopy was probably instrumental in providing the "good" conditions on earth before the flood. The vapor canopy could have blocked harmful radiation from the sun and contributed to the long lifespan of humans and animals. A vapor canopy would also have provided a greenhouse effect on the earth, providing a uniformly warm climate. This would allow for a widespread proliferation of vegetation from the north pole to the south pole. There is in fact fossil evidence of widespread tropical plants and animals from pole to pole; however, they are now in ice.

The vapor canopy was similar to our ozone layer in that it filtered out harmful radiation. This explains why people and probably animals lived longer. There would have been no harmful mutations caused by radiation and therefore animals and man would not have cancer and many other disorders which we now know come from genetic mutations. The ages of the men reported in the pre-flood genealogies are in the 800s and 900s. For example, Seth lived to be 912 years old. Enosh died at 905. Kenan was 910 when he died. Mahalalel died at 895, but his son, Jared lived to be 962. Harmful radiation may cause us to grow old quickly at present because the genes we need for everyday functioning could be broken down by radiation.

There were large plants and animals in pre-flood times. The greenhouse effect of the vapor canopy would allow for tremendous growth potential for organisms. Examples of large plants and animals found in the fossil record include:

Moss three feet tall
Ferns 50 to 70 feet tall
Giant Club Moss 100 feet tall
Dragonflies with 27 inch wingspans
Cockroaches one foot in diameter
Dinosaur — Very large!
Storks 20 feet long
Beaver 7 feet 6 inches long

If there was a vapor canopy covering the whole earth, there would be no extremes in temperature on the earth. There would be a uniform warm temperature all over the earth from pole to pole. There would be no high and low pressures because temperature was the same. Differences in temperature cause high and low pressures and weather such as hurricanes, etc. There would have been no rain and no weather as we know it. The fossil record shows tropical plants and large animals from pole to pole. The best explanation for this is that the temperature was warm all over the world which is what the vapor canopy theory would predict. There are numerous examples of fossilized tropical vegetation in now frozen arctic areas (e.g. Northern Canada-Palm trees in basalt; New Siberian Islands-fruit trees 20 feet tall).

As suggested by the scriptures in Genesis 7:11-12, the vapor canopy collapsed as one of the water contributions to the vast floodwaters necessary to cover the earth. The second source was water from beneath the earth.

11 In the six hundredth year of Noah's life, on the seventeenth day of the second month —on that day all the springs of the great deep burst forth, and the floodgates of the heavens were opened. 12 And rain fell on the earth forty days and forty nights.

There certainly would be a lot of water contained in a vapor canopy to help flood the earth quickly. Before we look at the results of the worldwide flood further, we need to examine some additional scriptures that help define the vapor canopy and its implications.

Prior to the flood, there was no rain. What does this say about a vapor canopy?

...and no shrub of the field had yet appeared on the earth, and no plant of the field had yet sprung up, for the LORD God had not sent rain on the earth and there was no man to work the ground, but streams came up from the earth and watered the whole surface of the ground (Genesis 2: 5-6)

No rain fell, but instead streams (also translated mist) were the means of watering the earth at this time. It implies that while the vapor canopy was intact, there wasn't rain. This fits very nicely with the concept of a greenhouse effect which preserved water within an enclosed system, thus reducing the evaporation. When the vapor canopy collapsed, at the flood of Noah, it ceased to exist. It wasn't repaired. Some of the changes that would be expected (post Flood) include the polar cold conditions and the rapid freezing of much of the floodwaters that were slowly receding. With the vapor canopy collapsed, the uniformly warm conditions would have ended and ice caps formed at the North and South poles. It is possible that ice deposition occurred rapidly and ice caps formed — what geologists call the Ice Age.

The Genesis account, just after the Flood, gives some outline to the concept of seasonal weather patterns versus the uniformity of a pre-Flood (pre canopy collapse) world.

While the earth remains, seedtime and harvest, and cold and heat, and summer and winter, and day and night shall not cease (Genesis 8: 22)

This is a description of the four seasons which, prior to the fall of vapor canopy, would not have existed.

Finally, it was apparent that no one had seen a rainbow prior to the Flood. In Genesis 9:13 God made a covenant between Himself and the earth that never again will water become a flood to destroy all flesh. He said "I set my bow in the cloud." As we all know, a rainbow requires the sun to shine and cast its rays on the clouds. As suggested earlier, the vapor canopy blocked the direct sun light and also, there was neither rain nor cloud.

The Flood

There are many models, from a creation science viewpoint that attempt to explain the events of the Flood. These models are part of what is collectively referred to as the flood model or catastrophism. Much of the geological features seen on the earth today are directly attributable to the catastrophic events of the Flood. There are certainly gradual geological activities, as well as episodic catastrophes that also constantly change the geology as well. Creation science doesn't dismiss these accepted scientific geological phenomena. Creation parts company with the uniformitarian view (evolution model) on the disagreement about the reality of a worldwide catastrophic flood. The Flood of Noah changed the climatic conditions of the earth and provided majority of the geological strata of the earth's crust.

The creation model explains the layers of the earth as being formed by the flood. After the flood, water had to drain off and into the oceans so that land could once more be seen. It is possible that the displaced water (cubic miles of water) formed tidal waves and caused tremendous erosional forces. The receding water would have been powerful enough to have formed canyons such as the Grand Canyon and also river beds. It would also result in vast amounts of sedimentary rock everywhere on the earth — even beneath the surface of bodies of water. This turns out to be the case.

Most notably, all of the life on the earth, except what was on the ark, was utterly destroyed. If you consider the tumultuous action of the flood waters, the death of the organisms, and the rapid silting over of the remains, you would expect to find a vast array of fossilized creatures in the geologic strata around the world. The fossil record does in fact support this suggestion. If you consider the types of organisms and their location, it appears that there is a relationship between two factors. First, a fossilized organism that was a victim of the flood, is found in its environmental niche. For example, fossil trilobites are often found in the lowest of geological strata. This isn't a surprise, since trilobites (like their modern counterparts the horseshoe crab) could be found in the sea crawling around on the bottom. This holds true for other examples, as well, that we will examine later. The second factor influencing the location of fossilized remains of Flood victims is based on the organism's ability to escape (at least temporarily) the raging and rising flood waters. Therefore, these fossils would be found at higher strata. They would also be rarer, since they might have escaped the silting process in many instances and been consumed by decomposing processes.

> 23 Every living thing on the face of the earth was wiped out; men and animals and the creatures that move along the ground and the birds of the air were wiped from the earth. Only Noah was left, and those with him in the ark.
> (Genesis 7: 23)

God instructed Noah to take representatives from each "kind" onto the Ark. Except for the fish in the sea, which are apparently excluded from the account of destruction, the only animals that survived this catastrophe were passengers aboard Noah's Ark. Much has been written on the plausibility of an ark carrying these representatives. Without going into all the detailed calculations, the ark was more than adequately large for all of its cargo. It is important to note that not all existing varieties were necessary to be brought aboard, only the species. So, using our dog example, there weren't 150 different breeds (or varieties) of dog brought onto the ark. Since all dogs are one species (or kind), then only representatives were required. The presumption here is that God would have directed which animals would be the ones taken so as to insure the genetic material for varieties was available when the earth would again be repopulated. There may not have been a great amount of variation in organisms pre-Flood, since the environmental conditions of the earth were likely uniformly warm in climate and may have been fairly homogeneous geographically.

It is difficult to say with certainty how the genetic material was preserved by the representative animals on the ark. This makes for an interesting debate, but doesn't detract from the logical nature of a supernatural event. In addition, God was in all likelihood directing the migration of the animals themselves to the location of the ark. This was supernatural, but possibly involves a mechanism that we know today as a migratory behavior.

What came off of the ark was a genetic bank of sorts. In other words, each representative species contained the genetic repertoire necessary to bring about the varieties to the earth that we see today or could see. It is unlikely that the type of variation we see today (*e.g.,* 150 varieties of dog) was present before the flood. In fact, I doubt if there was very much variation at all. For one reason, there wasn't much if any environmental pressure that would create the need for competition for a niche before the flood. Remember, the world wasn't perfect, but it still had the vapor canopy, so the climate was uniform.

Conclusion

The world and the universe around us provides boundless opportunity for study. This can begin with simple observation and lead to complex scientific analysis. It can involve abstract concepts and ideas or concrete facts. The model of scientific creationism ought to seek explanations for origin and diversity of life using the objective tools of the science disciplines and be subject to the types of rigorous scrutiny that any scientific process must endure.

Creation science is often criticized as being unscientific and driven by an evangelical agenda. Although many creation scientists do share a compelling sense of urgency about revealing the Truth, this shouldn't distract them from developing honest objective arguments.

The same should be true of the evolutionists who, on more than one occasion, have given in to philosophical persuasion. There are a number of evolutionists that unfortunately have been motivated by their rejection of intelligent design.

Creation science does operate under the caveat of intelligent design as the *sine qua non* versus chance. It is the boundaries of the Bible that define the creation model.

Part Three: Examining the Evidence

CHAPTER NINE

How to Test Evolution and Creation

Part Two examined the foundations and frameworks for both evolution and creation. Creation and evolution have diametrically opposed views with regard to the origin and diversity of life. The concepts of evolution are rooted in philosophical speculations, based entirely on natural explanations, and random chance events. The concepts of creation incorporate true scientific principles, but accept supernatural explanations, and purposeful intelligent design.

In the process of developing and discussing the foundational material in Part Two of this book, several key assumptions and predictions were identified which define the premises for creation and evolution. Care was taken not to evaluate these assumptions and predictions extensively in those earlier chapters. In this chapter, and in subsequent chapters, we will examine the observable evidence to see how well creation and evolution are supported. First, we need to establish the approach and criteria for evaluation.

Much is written in both popular and scientific literature concerning evolution and the so-called "theory" of evolution. We need to understand what a theory is before we can apply it appropriately to either evolution or creation. The word *theory*, as used in the scientific world, has a little bit different meaning than when it is used in everyday English language. Usually theory in everyday conversation has about the same meaning as *idea* — such as "I have a theory about how our neighbor's house burned down — it might have something to do with their child loving to play with matches." In scientific language, however, a theory is a very well tested hypothesis and it is not just a statement or opinion such as "vitamins can cure cancer." Theories are broad generalizations about how the universe works including the biological world. Some examples of currently accepted theories in biology include:

- **Cell Theory**. All organisms are composed of one or more cells with the exception of viruses which are only made of a protein coat and either DNA or RNA. The cell is the structural unit of all organisms. All cells come from preexisting cells.

- **Biogenesis Theory**. Life comes only from life. In other words no person or scientist has ever created life, even small forms such as bacteria.

- **DNA/RNA is the Blueprint for Life**. All life contains DNA/RNA (found in chromosomes) which is coded information that dictates or controls what the animal/plant will be like: its structure or anatomy, its function or physiology, and its behavior.

A theory provides a good explanation for scientific observations. Perhaps most importantly, a theory has been tested by many scientists over a long period of time. It has withstood the test of time and found to be true within the realm of our present human knowledge. A theory is also a good tool for predicting what will happen in the physical and biological world.

Sometimes scientists and laymen may even think and talk about theories as if they were facts; however, a theory can still be proved wrong. In 140 AD a Greek astronomer Ptolemy put forth the geocentric theory which stated that the sun revolves around the earth. It was not until 1530 AD that Copernicus finished his heliocentric theory that the earth revolves around the sun.

It is easy to see that treating creation or evolution as theories would be difficult due to many semantic reasons. From a scientific perspective, it is difficult as well since no one was present at either the beginning of first life or at the beginning of new species, so there is no observations to serve as the basis for a theory. In the absence of observations, any initial hypothesis would therefore have to be based on someone's philosophical idea or faith in a Biblical account. Subsequently, testing a hypothesis of this nature wouldn't prove anything. Trying to approach creation or evolution as theories won't work. It would be like trying to explain the origin of water by turning on the water faucet. However, there is a scientific approach that can be applied correctly to our evolution/creation problem — the model approach.

Models, on the other hand, are abstract representations that establish frameworks within which analysis of problems can occur. Both creation and evolution have outlined this framework in their numerous models and submodels. Here's one example: In general, Darwinian evolution explains the formation of new species by an accumulation of favored variations over a very long period of time. This identifies a couple of variables for the evolution model. One is *time* and the suggestion is — a very long time due to the vast amount of time required for evolution to account for the diversity of life we know. A second variable is evidence of the transitional forms (intermediates showing the succession of variations). So although we can't test evolution as a theory, we can make observations regarding how well the known values fit the model. If the observable evidence doesn't fit, then the model is wrong and must be changed or abandoned.

We should now be satisfied with calling evolution and creation both models for the purpose of evaluating them. We will look at each model individually in three categories: origin of life, diversity of life, and age of the earth. Some salient predictions will be presented in this chapter, then evaluated in the subsequent chapters.

Table 9-1 represents the variables and predictions for both the creation and the evolution models. For each prediction, we will examine the evidence to see how well each model holds up to objective evaluation. The literature on both creation and evolution is very large and full of data and interpretation. Some of it is very interesting and enlightening; however, it is beyond the scope of this book to review all the literature. We will examine some historically important efforts and the salient current arguments as well.

The next three chapters will address the predictions in the three categories of variables: Origin of Life (Chapter 10), Diversity of Life (Chapter 11), and Age of the Earth (Chapter 12). It is not the intent of this book to thoroughly review each and every value proposed for the predictions identified in Table 9-1. Furthermore, for various reasons, we can't. First of all, the number of variable values is much too large. We have chosen some of the more salient and illustrative ones for discussion. Second, each of the values is based on years of work, and we can't duplicate that level of work for each variable value. There has been a great deal of reexamination done by noted scientists, but this only represents the controversies. Furthermore, the claims surrounding the controversies can't be resolved by literature review. For every claim there is an equal and opposite rebuttal from one side of the argument or the other. In many cases, resolution of the arguments would require reexamination of fossils or conducting experiments, which is impractical and impossible for us to do as part of writing this book. Even reexamination would only provide another opinion, for which there are already plenty on both sides.

	Model	Variable	Predictions
Origin of Life	Evolution	Life originated by the chance assemblage of chemical components	- Can produce life in lab. - Might observe spontaneous generation of life in nature.
	Creation	Life was created as an special intentional act by an intelligent designer	- Can't produce life in lab. - Will not observe spontaneous generation of life in nature.
Diversity of Life	Evolution	Natural selective pressures can favor certain variations that will form new species	- Transitional forms that represent the intermediates will be found as living examples. - Transitional forms that represent the intermediates will be found as fossilized examples.
	Evolution	Transitions take place in local populations, then spread rapidly as a new species	- Transitional forms that represent the intermediates will be found as fossilized examples in localized geographic areas.
	Creation	Natural selective pressure can favor variations	- Shifts in gene frequency will be observed in populations of species exposed to changes in environment.
	Creation	Kinds or species were created within the six days of Biblical creation	- Transitional forms that represent the intermediates will NOT be found as living examples. Transitional forms that represent the intermediates will NOT be found as fossilized examples.
Age of the Earth	Evolution	The earth is somewhere around 4.5 billion years old.	- Various chronographic measurements should support an old earth.
	Creation	The earth is relatively young (less than 10,000 years.	- Various chronographic measurements should support a young earth.

Table 9-1. Predictions for Origin of Life, Diversity of Life, and Age of the Earth Variables

The literature is replete with the descriptions, commentary, and dispute. There are some very thorough and well-documented work that provides "evidence" contradicting or supporting values for the variables in both the creation and evolution models.

What we will attempt to do in the next three chapters is present a fair review of the variable values and provide some of the criticisms of those values. Once again, this won't be a thorough review of the literature. If the reader wishes to challenge any of the variables, they are encouraged to do so on their own. However, we have found that relying on someone else's opinion isn't a strictly reliable method of resolution. There will always be an educated list of proponents with expert judgement and opinion for every controversy.

CHAPTER TEN

The Origin of Life

The experimental and analytical work in this area has been unsuccessful in producing life, even in its simplest form. There have been some interesting and enlightening experiments that have predictably produced various chemical components of life under highly manipulated conditions.

In the six days of creation, we saw that God created all living things. The Bible didn't say that God created some things and then let nature take its course. It clearly says that each was created according to its kind. This clearly defines the origin of life. Surprisingly, even among some ardent evolutionists, there is an element of acquiescence when it come to the origin of life itself. Creation of first life is often attributed to God, or some supernatural force.

In this chapter, we will briefly review the history of the abiogenesis concept and its cycle of becoming fashionable and less fashionable. We will examine some of the notable exemplary origin of life models and see how well their mechanisms hold up. Finally, we will consider the fairly recent proposals that life originated on another planet, then was introduced to earth by meteorites or other vehicles from space.

Before we explore the origin of life proposals, let's look at the specific predictions for the origin of life variable in Table 10-1.

	Model	Variable	Predictions
Origin of Life	Evolution	Life originated by the chance assemblage of chemical components	- Can produce life in lab. - Might observe spontaneous generation of life in nature.
	Creation	Life was created as an special intentional act by an intelligent designer	- Can't produce life in lab. - Will not observe spontaneous generation of life in nature.

Table 10-1. Predictions for the Origin of Life Variables

Brief Review of the Abiogenesis Theories

Anaximander of Melitus was a Greek mathematician and astronomer who, around 550BC proposed that somehow, life arose from the slime in the sea. This was one of the earliest alternatives to creation as an explanation of how life may have originated. This concept, called *spontaneous generation*, was popular among some evolutionists until the mid-1800s. Spontaneous generation is also referred to as *abiogenesis* meaning originating without life. These abiogenic ideas were encouraged by the observation of flies "arising" from meat and rats emerging from piles of rags and garbage. Although experimentally disproved by the Italian physician Francesco Redi in 1668, these spontaneous generation ideas persisted. The discovery of the microscope and its revelation of the microbial world led many scientists to suggest that these "simple" organisms could be generated spontaneously.

Flawed experiments conducted by John Needham in 1745 provided convincing evidence of spontaneous generation. In his experiments, nutrient broth was heated, which killed the microbes. However, a short time later, organisms were again seen in the flasks. This brought spontaneous generation back in favor for twenty more years. Once again, a redesigned experiment by Lazarro Spallanzani correctly suggested that the microbes may be entering (*i.e.,* contaminating) the broth through the air. Lazarro suggested sealing the containers after the heating process. Sure enough, the microbes were not present in the heat-treated and sealed containers. Needham's response was that there was a "vital force" that was eliminated in the heat treatment and wasn't able to return due to seal on the containers. It was also suggested that the sealed containers eliminated the oxygen needed to support life.

In 1858, Rudolf Virchow proposed his theory of biogenesis, which proposed that life came only from life and that cells only came from other cells. It wasn't until the French scientist Louis Pasteur executed his experiments, that the issue was nearly resolved. Pasteur's experiments were able to account for the criticisms. He designed an apparatus in which the nutrient broth was heat-treated, thus killing the existing organisms. His apparatus was U-shaped which allowed the air (oxygen) to enter, but trapped the contaminating particles from entering the broth. For a few years following Pasteur's elegant dismissal of abiogenesis, there was little more said.

Twentieth Century Spontaneous Generation

To the modern scientist with an understanding of microbiology, these early concepts of abiogenesis seem almost laughable. The idea of spontaneous generation of rats from rags and flies from old food is ludicrous. However, for some reason, life spontaneously generating from a chemical soup isn't laughable. For the evolutionist, it is merely raising the bar of expectation based on a refined insight into the ultra structure of organisms. So the flies from garbage theory has turned into cells from primordial chemistry.

Let's begin our evaluation of origin of life by considering the currently proposed progression of earliest components to what may be considered life. The general progression would look something like Figure 10-1.

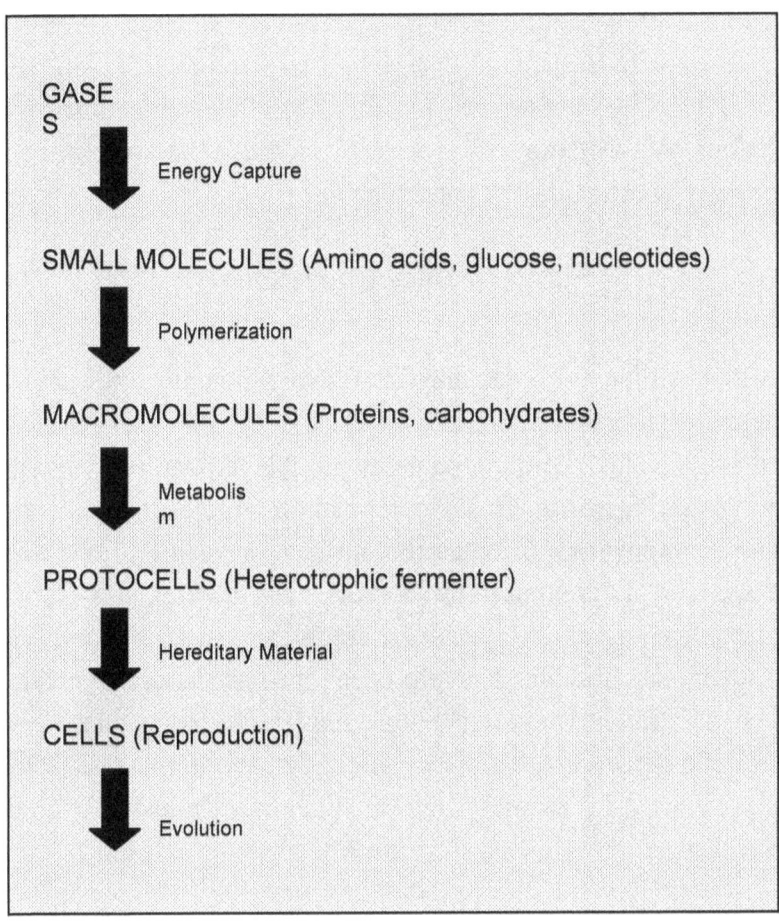

Figure 10-1. A Progression from Non-Life to Life (Adapted From Mader)[24]

One of the early leaders in this line of thought was a Russian scientist named Oparin. In 1924, he proposed a theory that certain primordial conditions and events could theoretically give rise to life. Promising work in this area was undertaken at the University of Chicago by a scientist named Stanley Miller in 1950s. His approach, based on Oparin's theory, was to try to determine the recipe for the

---

[24] Sylvia S. Mader, *Inquiry Into Life* (Dubuque: W. C. Brown, 1991), p. 553

primordial soup and then build an appropriate apparatus to produce life. Unfortunately, neither he nor any scientists since, have been able to come up with a successful experiment that produces life in any form. However, in fairness to our objective of evaluating the predictions of the models, we will examine the premises and outcomes of some of these experiments.

Stanley Miller Experiments

Before Miller could do anything experimentally, he needed to somehow establish the conditions of the earth's early primordial environment. This required considering the temperature, humidity, ultraviolet radiation, chemical composition of the atmosphere, and chemical composition of the earth's surface. We can tell you already, these are a lot of assumptions. Anyone who has taken high school chemistry knows that each of these conditions, in a test tube type of experiment, are critical. The presence or absence of any chemical (even in the smallest amount) can completely change the outcome. The slightest increase or decrease of temperature, can affect a chemical reaction. The post-conditions can also impact the products of the reaction as well. So, since no one knows what the primordial earth conditions were, the assumptions made by Miller (or anyone else) are educated guesses suggesting that the starting premises may be fatally flawed. The following quote from a standard college biology textbook will illustrate the evolution of thought in this area:

> The gases of the primitive atmosphere were not the same as those of today's atmosphere. Originally, it was proposed that that the earliest atmosphere contained a lot of hydrogen (H_2) because it is the most abundant element in our solar system. Later, it was suggested that lightweight atoms, including hydrogen, would have been lost as the earth formed because the earth's gravitational field was not strong

enough to hold them. Now it is thought that the primitive atmosphere was produced by outgassing from the interior, particularly by volcanic action, after the earth formed. If this was the case, the atmosphere consisted mostly of water vapor (H_2O), nitrogen N_2), and carbon monoxide (CO). [25]

Since Miller's ideas on what the early earth conditions were are as good as any, let's see how things turned out with the starting assumption that the early atmosphere contained methane — the starting materials for amino acids. Amino acids are molecules that, when connected in a specific sequence, form the proteins that make up living things. It is important to note, that amino acids are chemical compounds. Experimentally producing a chemical molecule isn't creating life. For example, burning gasoline in a car engine produces carbon dioxide and carbon monoxide. Living things produce those chemicals as well during respiration. Can we suggest that a car is living — No! Furthermore, the results of these experiments don't even start to explain how, once these chemical compounds were formed in a primordial organic soup, they were able to form into anything more significant.

Figure 10-2 is a highly schematic representation of the type of apparatus Stanley Miller used in his experiments.

---

[25] Mader, p. 553.

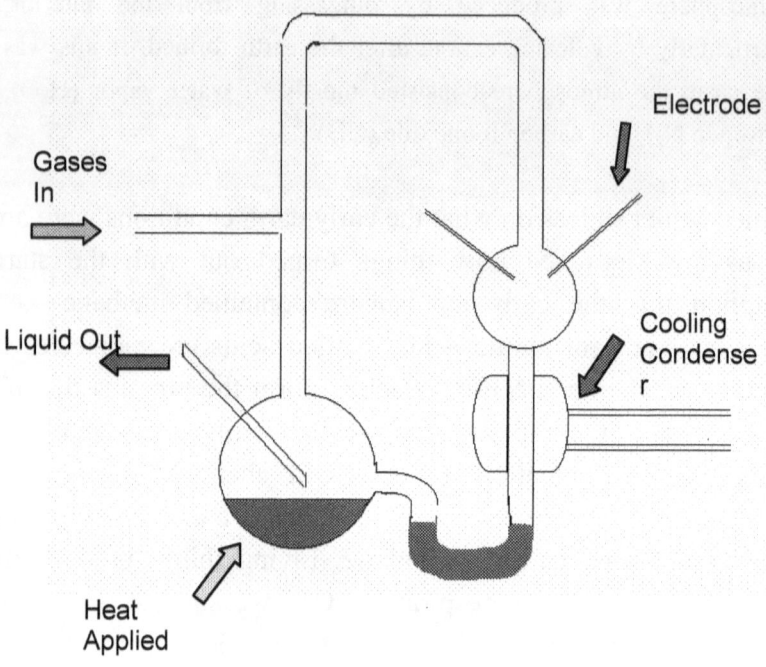

Figure 10-2. Miller's Origin of Life Apparatus. Adapted from Mader[26]

Gases were added to the sealed environment. Electrical sparking provided a simulated lightening strike as a catalyst for reaction, the gases were condensed by cooling, then the resulting liquid was kept boiling by heating the collecting flask. The liquid in the collecting flask was then withdrawn and analyzed. Sure enough, what resulted were amino acids — that's what chemistry does.

Subsequent similar experiments by Miller and others also produced amino acids. Michael Behe, a biochemist, points out that this, and all origin of life experiments, require intelligent manipulation of the mechanisms employed. This makes the results more of a forced lab protocol than a demonstration of life originating

---

[26] Mader, p. 555

in an abiotic earth environment. Behe and others have made a critical point that applies to all the origin of life research: "Joining many amino acids together to form a protein with a useful biological activity is a much more difficult chemical problem than forming amino acids in the first place."[27] Proteins are comprised of chains of amino acids. Not just any chains — one wrong amino acid and there is no protein. Behe points out that another of the main problems here is the requirement for removing a molecule of water for each amino acid that was added to a chain; and the presence of water inhibits protein formation. This is truly a dilemma given the proposed nature of the primordial earth — full of water.

Sidney Fox and Dry Heat Polymerization

This water problem gave rise to an idea, by Sidney Fox of the University of Miami, that amino acid polymerization (hopeful protein formation) would take place by exposing the amino acids to dry heat. His model suggests that the amino acids would collect in shallow pools, then when the water evaporated, they would be heated and the polymerization would occur. These molecules were called proteinoids. There was a great deal of manipulation as well with this model and its plausibility was questioned by both evolutionists and creationists alike.

Thomas Cech and RNA

In the 1980s, Thomas Cech at the University of Colorado and Sidney Altman at Yale, speculated that RNA was the first molecule to exist and not a protein. The attraction was that RNA could replicate and regulate itself. This was a dream come true for origin of life research and earned Cech and Altman the Nobel prize in 1989. Walter

---

[27] Michael J. Behe, *Darwin's Black Box* (New York: Touchstone, 1996), p. 169.

Gilbert, a Harvard biochemist, also believed in 1986 that the first "organisms" were comprised of RNA molecules and evolved to be able to synthesize the proteins, etc. [28]

Michael Behe's response to the RNA view:

> The big problem is that each nucleotide "building block" is itself built up from several components, and the processes that form the components are chemically incompatible. Although a chemist can make nucleotides with ease in a laboratory by synthesizing the components separately, purifying them, and then recombining the components to react with each other, undirected chemical reactions overwhelmingly produce undesired products and shapeless goop on the bottom of the test tube. [29]

Dr. Leslie Orgel, an origin of life researcher at the Salk Institute, said that "Experiments simulating the early stages of the RNA world are too complicated to represent plausible scenarios for the origin of life."[30] In a Scientific American review article, John Horgan indicates that Orgel is not alone in his skepticism. Horgan suggests that many researchers recognize the problems with the RNA proposal as well: "RNA and its components are difficult to synthesize in a laboratory under the best of conditions, much less under plausible prebiotic ones."[31]

Life Originating on Mars

As we have seen, the model and experiments attempting to provide evidence of life originating spontaneously and evolving

---

[28] John Horgan, "In the Beginning," *Scientific American*, February, 1991, p. 119.
[29] Behe, p. 171.

[30] Horgan, p. 119.
[31] Horgan, p. 119.

chemically have been far less than successful. So, there's always the possibility that life actually originated somewhere else (*i.e.,* on another planet) and seeded the primordial earth — possibly riding a *meteor*. If this seems far-fetched, read on. This is exactly what was considered in 1996.

The story actually began in 1984 with a team of explorers in Antarctica who discovered and recovered a number of rocks which were subsequently identified as being fragments of a meteorite. This meteorite is presumed to have resulted from a collision between Mars and something else 16 million years ago. The resulting fragments then floated in space until falling to Earth 13,000 years ago. Among the rocks, one in particular, so-called Alan Hills (or ALH) 84001 was the subject of some interesting claims. Twelve years after the discovery, NASA announced in August 1996 that they believe that the meteorite showed signs that ancient life existed on Mars. Various lead stories went out in page one news such as "MARTIAN MESSAGE-THERE WAS ONCE ANCIENT LIFE HERE"[32] Network and cable television were also running these stories. Here is the basis for these claims.

The NASA scientists leading the research, Dr. David McKay and Dr. Everett Gibson, largely base their arguments on the existence of two things found in the ALH 84001 rock. The first was what appeared to be polycyclic aromatic hydrocarbons (PAH) with traces of magnetite and iron sulfide with the PAHs. These chemicals can be associated with biological activity. On earth, PAHs are known to be formed by bacteria. The second evidence is what the researchers said were fossilized nanobacteria (*i.e.,* very small bacteria) which were so small that an electron microscope was required to visualize them. At a meeting of the Planetary Division of the American Astronomical Society, the NASA team (Dr. McKay and Dr. Gibson) received what

---

[32] Robert C. Cowen, "Martian Message-There was Once Ancient Life Here," *Christian Science Monitor*, 8 August 1996, Col. 1, p. 1. (Proquest Abstract 04170423).

was called "blunt skepticism" from other scientists.[33] When pressed on their evidence, several times Dr. McKay and Dr. Gibson admitted that they hadn't yet developed enough evidence to settle the controversies. Subsequent articles refer to reports that the so-called microfossils weren't nanobacteria at all, but were magnetite crystals formed from high-temperature vapor processes. Other articles report explanations arguing that the findings are similar to those found in volcanic vents and unlike those formed in biological activity. Contradictory evidence from scientists at the Georgia Institute of Technology suggested the magnetite crystals "contain structural defects that are not known to be produced in a biological environment."[34]

Conclusion

It seems apparent that life has not been produced in the lab, nor has it been discovered originating spontaneously in nature. Even the attempts at producing building blocks with forced laboratory protocols are wrought with complications that are only resolved with intelligent manipulation. Research continues, but this quote from Klause Dose provides some insight into the level of optimism for the origin of life researcher:

> More than 30 years of experimentation on the origin of life in the fields of chemical and molecular evolution have led to a better perception of the immensity of the problem of the origin of life on Earth rather than to its solution. At present all discussions on

---

[33] "NASA Experts Defend Theory of Life on Mars," *New York Times*, 25 October 1996, p. 24 (ProQuest 9300159612).
[34] Robert Cowen, "More Findings About Life on the Red Planet," *Science News*, 8 February, 1997, Vol. 151, p. 87.

principle theories and experiments in the field either end in stalemate or in a confession of ignorance.[35]

The search for origin of life goes on unabated. The director of NASA's Astrobiology Institute (NAI), Baruch Blumberg, is actively pursuing the origin of life research. A Scientific American profile in July 2000 reports that the "NAI now comprises some 430 astrobiologists at 11 universities and research institutions."[36]

---

[35] Behe, p. 168.
[36] Julie Wakefield, "The Search for Extreme Life," *Scientific American*, July 2000, p. 30.

CHAPTER ELEVEN

The Diversity of Life

The second set of variables that we will consider are the diversity of life variables and the associated predictions. If, as Darwin believed, all life is related to a common ancestor, it is very reasonable to expect that over time the displaced transitional forms should be found in abundance in the fossil record. Under this evolutionary model, it would also be reasonable to expect that some number of transitional forms may still be alive as displaced and doomed predecessors of the new species. If, on the other hand, the creation model holds true, the fossil record should reflect distinct representations of species many of which are extinct. The creation model also predicts a shift in gene frequency that results from environmental pressure favoring a particular trait or traits in a population of organisms. This is an idea that is shared with the evolutionists and is one that can be observed (even tested) to an extent in nature and in the laboratory.

In this chapter we will examine some of the purported discoveries of transitional forms. Bear in mind, the literature is full of paleontological material, much too voluminous to review for this book. The frenzy to discover a missing link has sent anthropologists, paleontologists, archaeologists, biologists, graduate students, and even amateur fossil collectors on a relentless hunt. Even the noted paleontologist Tim White is critical of scientists' tendencies to name something as a new species based on finding fossils with slightly different morphologies. This chapter will only examine some of the salient nominations and lead the reader to be vigilant and objective in evaluating other reports.

The mechanisms for speciation (one species transitioning to another) was discussed in a previous chapter. There are only two explanations that persist among the evolutionists: One is *gradual evolution* and the second is *punctuated equilibria*. Both explain the arrival of a new *Species B* as resulting from the accumulation of favorable mutations in *Species A* over time. In other words, the pool of genes available in a population of *Species A* changed over time due to mutations. These mutations found their way to the population by varying mechanisms. The mutations continued to accumulate as the environment favored these changes to the point where *Species A* (*i.e.,* its genetic identity) was lost in favor of the new *Species B* (*i.e.,* its genetic identity). In this way, evolution explains how a so-called primitive amphibian could over time become a frog and, over vastly more time, become a dog. *Punctuated equilibria* differs in that the transitions occur in short bursts in geographically isolated regions, then they spread. Regardless, for either of the evolutionary approaches, there needs to be evidence to support the transition, either fossil evidence or living examples. One way to view this graphically is through a form of tree and branches. Using the previous speciation example (*i.e.,* Species A becoming Species B), a branch could be viewed in the following way.

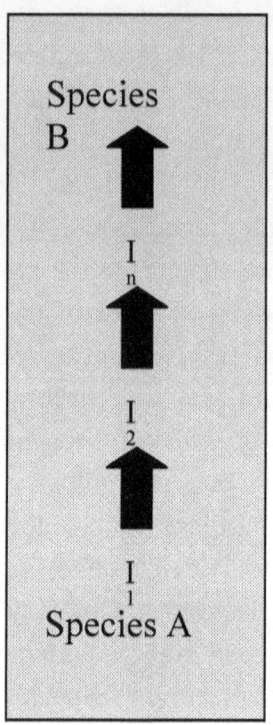

Figure 11-1. Example Sequence Illustration[37]

The intermediates (I^n) should be numerous and represent the transitional forms. Each intermediate should show a continuity from one to another. In other words, if the intermediates exist, the Intermediate (I^2) will exhibit an extremely minute difference from its predecessor (I^1). This difference would represent the mutation(s) and their effect on the species morphology — presumably a morphology that is favored and accumulated with the mutation "enjoyed" by its successors.

Location and identification of the intermediates (*i.e.,* the I^1, I^2,...I^n) has been a passionate endeavor by evolutionists since the time

---

[37] Format taken from Michael Denton, *Evolution: A Theory in Crisis* (Bethesda, Maryland: Adler and Adler, 1986).

of Darwin and the literature is full of reports and claims, the most important of which we will examine in this chapter.

The creation model acknowledges the shifting of gene frequency in populations of species, but doesn't expect to see transitions from one species to another. The expectation is to see a sudden appearance of all forms of life (explained by a creative act), variation within a species, and even extinction when environmental pressures exceeded the populations genetic potential to survive.

Before getting started, let's review the Diversity of Life variables and the predictions summarized in Table 11-1.

	Model	Variable	Predictions
Diversity of Life	Evolution	Natural selective pressures can favor certain variations that will form new species	- Transitional forms that represent the intermediates will be found as living examples. - Transitional forms that represent the intermediates will be found as fossilized examples.
	Evolution	Transitions take place in local populations, then spread rapidly as a new species	- Transitional forms that represent the intermediates will be found as fossilized examples in localized geographic areas.
	Creation	Natural selective pressure can favor variations	- Shifts in gene frequency will be observed in populations of species exposed to changes in environment.
	Creation	Kinds or species were created within the six days of Biblical creation	- Transitional forms that represent the intermediates will NOT be found as living examples. Transitional forms that represent the intermediates will NOT be found as fossilized examples.

Table 11-1. Predictions for Diversity of Life Variables

With either *gradual evolution* or *punctuated equilibria*, the predictions would include the existence of transitional forms. In order for either to be supported, there must be evidence that satisfies those predictions. The creation model would predict exactly the opposite — there should be no transitional or intermediate forms found, either in a localized geographical region or in global geography. The creation model does acknowledge variation within a species and expects to see that type of variation throughout the fossil record and among living organisms today. However, the links between species will be missing.

Living Transitional Forms

This is the shortest of the discussions on transitional forms because there are none. The earnest search for these living transitional forms began at the time of Darwin with the high hopes of locating a missing link. Over the years, there have been numerous discoveries of new and often unusual life forms, but none that qualify as transitional forms. We are not aware of any viable nominations presently being considered by evolutionists. The possibility is still entertained by the evolutionists since living transitional forms are within the realm of expectation; however, there are none. This evolution prediction is therefore unfulfilled.

Fossil Transitional Forms

Darwin held much hope that the fossil record would bear out his theory. It was at that time that a flurry of paleontological activity began and has gone unabated to this day. The objective of finding a transitional form has motivated evolutionists in search of support for this prediction. The good news here is that this urgent paleontological investigation has provided some excellent results regarding an appreciation and understanding of the morphology of organisms that lived long ago. However, has the search and study of fossils revealed anything that would support the predictions about the speciation and the diversity of life?

We will divide the nominations into two categories for the purpose of analysis: The non-human transitional form nominations and the human transitional form nominations. The hypothetical human transitional forms will include the nominations for bridging the morphological gap between the human and the apes. The non-human transitional form nominations can include the hypothetical transition between any organism (*e.g.,* between a frog and a fish).

Non-Human Transitional Form Nominations

This may be a surprise to the reader, but there are only a couple of nominations that have held any interest.

Archaeopteryx

We will begin with one of the most notable nominations and one of the earliest — *Archaeopteryx*. Since 1861, there have been seven fossil specimens of *Archaeopteryx* found in Germany. The first and second specimens were identified in 1861 and 1877. The other five were identified after 1950. For the evolutionist, *Archaeopteryx* continues to represent a transition or intermediate between the reptile and the bird. Examination reveals that it had definite feathers, which is the distinguishing feature of all birds. It also has an opposable large toe (or hallux) which is another avian or bird characteristic. What appears to be different in *Archaeopteryx* are some skeletal features, such as a breast bone that was, in some ways, reptilian in form (*i.e.,* flat). However, many species of bird, such as ostriches, have similar features (*i.e.,* flat breastbones). *Archaeopteryx* also has teeth and three clawed fingers at the end of its wings; however, according to Charig (an ardent evolutionist) "the classic *Archaeopteryx* is surrounded by the impressions of unmistakable feathers and is therefore classified quite positively as a bird."[38] He assesses *Archaeopteryx* as the oldest fossil bird and pronounces that it is NOT a reptile-bird intermediate.

Much debate continues on *Archaeopteryx* as a transitional form. Denton suggests that "ninety-nine percent of the biology of any organism resides in its soft anatomy, which is inaccessible in a fossil."[39] The skull of *Archaeopteryx* however reveals "its brain was essentially avian in all important respects...The possession of an essentially avian central nervous system lends further support to the idea...that *Archaeopteryx* was as capable of powered flight as a typical modern bird" and therefore would have had all the other associated avian anatomical components as well.[40] Therefore, this single nomination provides little convincing evidence to support it as a transitional form.

Longisquama

Other promising nominations have made the news fairly recently. These are so-called dinosaur-bird transitional forms. The first was a specimen, named *Longisquama*, discovered in Russia in 1969. In 1999, it was on exhibit in a shopping mall in Kansas City, Missouri. A research team believed it had what appeared to be feathers. However, this assessment has received criticism from other paleontologists. Jacques Gauthier, a dinosaur evolution expert at Yale provided earlier criticism in a June 2000 *USA Today* article calling *Longisquama* a poorly preserved specimen and important only "if you

---

[38] A. Charig, *A New Look at Dinosaurs* (Facts on File, Inc., 1988), p. 83.

[39] Denton, p. 177.
[40] Denton, p. 177-178.

allow your imagination to run wild."[41] A news brief appearing in *Scientific American* also reports that paleontologists have provided criticism of the featherlike structures arguing "that although the structures are indeed unique, they are probably scales, not feathers."[42]

Protoarchaeopteryx robusta and Caudipteryx zoui

Protoarchaeopteryx robusta and *Caudipteryx zoui* represent nominations that are so-called feathered dinosaurs from China's Liaoning province — *Protoarchaeopteryx robusta* and *Caudipteryx zoui.* The discovery and analysis appeared in a *Nature* article in 1998, describing the morphology of a specimen named *Protoarchaeopteryx robusta* and two specimens of *Caudipteryx zoui,* proposing these two dinosaurs as having features that are consistent with theropod dinosaurs and some similarities to *Archaeopteryx.* Noted in the *Protoarchaeopteryx robusta* were the presence of plumulaceous feathers anterior to the chest, associated with caudal vertebrae, along the lateral side of the right femur and the proximal end of the left femur. The *Caudipteryx zoui* specimens show the presence of body, wing, and tail-type feathers.[43]

The CNN online headline read "Scientists: Fossils prove that birds evolved from dinosaurs." "They represent a missing link between dinosaurs and birds which we had expected to find," said Ji Quiang, director of the National Geological Museum in Beijing, who worked on the fossils.[44] However, like many of the nominations, this one isn't free of controversy. Soon after the *Nature* article, the questions and criticisms began to come. In the same CNN article, "Alan Feduccia,

---

[41] "Fossil Challenges Theory of Origin of Birds," *USA Today Online*, 22 June 2000, www.usatoday.com.

[42] "Down With Dino Birds," *Scientific American*, September 2000, p. 32.

[43] Ji Qiang, Phillip J. Currie, Mark A. Norell, and Ji Shu-An, "Two Feathered Dinosaurs from Northeastern China," *Nature*, Vol. 393, 25 June 1998, pp. 753-761.

[44] "Scientists: Fossils prove that birds evolved from dinosaurs," CNN.com SCI-TECH, June 24, 1998 Web posted at: 12:32 a.m. EDT (0432 GMT).

an evolutionary biologist at the University of North Carolina, Chapel Hill, said the discoveries are 'very interesting,' but he said they do not provide immediate and final proof that birds evolved from dinosaurs." Feduccia went on to suggest that the fossils could be primitive birds that happened to resemble dinosaurs.

Archaeoraptor liaoningensis

This was another of the "feathered dinosaurs" discoveries found in the Liaoning province in China. The find was presented at press conference given by Philip J. Currie of the Royal Tyrell Museum of Paleontology in Alberta, Stephen Czerkas of the Dinosaur Museum in Blanding, Utah, and Xing Xu of the Institute of Vertebrate Paleontology and Paleoanthropology in Beijing. The November 1999 issue of *National Geographic* magazine presents *Archaeoraptor* to the general public with a subtitle "NEW BIRDLIKE FOSSILS ARE MISSING LINKS IN DINOSAUR EVOLUTION,"[45] The study and analysis of the specimen was led by Stephen Czerkas who is quoted as saying "This fossil is perhaps the best evidence since *Archaeopteryx* that birds did, in fact, evolve from certain types of carnivorous dinosaurs."[46]

Many paleontologists and bird experts questioned the existence of the reported feathers and suggested the fossil may have been pieced together and might represent more than one animal. The authenticity of the find became increasingly questionable and support was soon withdrawn. Philip Currie said "It's the craziest thing I've ever been involved with in my career."[47] The *Science News* article also reveals that there are a number irregularities such as bones missing

---

[45] Christopher P. Sloan, "Feathers for T. Rex," *National Geographic,* November 1999, p. 99.
[46] Sloan, p. 101.
[47] R. Monastersky, "All Mixed Up Over Birds and Dinosaurs," *Science News,* Vol. 157, No. 3, 15 January 2000, p. 38.

connecting the tail to the body, which are a sign of "reworking." Storrs Olsen of the Smithsonian Institution's Museum of Natural History said of *Archaeoraptor,* "There probably has never been a fossil with a sadder history than this one."[48]

First Nominations for Human Transitional Forms

In other chapters, we have explained the Linnean taxonomic classification system used to study and organize living things. The only member of the family Hominidae is the human. The evolutionists suggest that there were other members of this family, which are now extinct. There have been a few claims, notably so-called missing links or transitional forms (the so-called ape-men) that allegedly preceded modern humans. This followed Darwin's notion that all living things have ancestral relationships with more primitive forms. Hypothetically, organisms, including humans, were descended with modification as a result of the natural pressures exerted over time and thus leaving behind the remains of the predecessor intermediates. This sparked the imaginations of many, inspiring them to seek the so-called missing links — especially hypothesized missing links between man and ape.

What follows in this section are accounts of some attempts to be first to discover that transitional form. As we will see, these early attempts failed. Sadly, even after some of these discoveries were analyzed revealing the errors, there was a persistence in the minds of some that these discoveries were somehow true.

Neanderthal

Today we know Neanderthal (or Neandertal) was a fully modern human. However, Neanderthal is an example of the first attempt at

---

[48] Monastersky, p. 38.

producing and exploiting a missing link or transitional form between the human and the ape. In 1856, the first Neanderthal skeleton was found in the Neander Valley — a limestone gorge with the Dussel River running through it — near the village of Hochdal, Germany. Since that time, numerous other Neanderthal bones have been found in many countries on three continents (Iraq, China, Israel, Hungary, Greece, Central and North Africa).

Even in its early exploitation, in 1857, it was concluded that Neanderthal represented a human, despite the different characteristics (*e.g.,* prominent brow). It was suggested that the Neanderthals suffered from a vitamin D deficiency that led to the disease - Rickets. However, this may not have been the case. It is entirely possible that Neanderthal was merely exhibiting variation in morphology.

In the early 1900s, Marcellin Boule of the National Museum of Natural History in Paris issued a scholarly paper that presented Neanderthal as ape-like which pleased the evolutionists. However, when critically examined, the following facts were found: The cranial capacity of the Neanderthal was 1600 cubic centimeters. This was larger than a modern human cranium, which averages between 1450 and 1500 cubic centimeters. In 1957, the bones were reexamined by Strauss and Cave, anatomists at St. Bartholomew's Hospital Medical College (London). They determined that the bones, especially the foot, were artificially positioned to give the appearance of being ape-like with a prehensile (opposable) large toe. Also, the assessment of a diseased condition was also substantiated. It is now recognized by all scientists that Neanderthal was in fact fully human and not ape-like. Furthermore, there is evidence that Neanderthal had a complex society and religion.[49]

---

[49] M. Pitman, *Adam and Evolution* (London: Rider and Company, 1984), pp. 86-89.

Java Man

Java Man (*Pithecanthropus*) was the next nomination for a transitional species. The existence of an ape-man was actually imagined by the German philosopher Haeckel in the 1880's and thought that it might be found in Southern Asia or Africa. He even commissioned an artist to paint a picture of the imagined missing link. Haeckel's student in medical school, Eugene DuBois, went looking for it in the Dutch West Indies as a member of the Royal Dutch East Indies Army. After two years of searching in Sumatra, he was transferred by the Dutch army to Java. In the fall of 1891 on the bank of the Solo River near the village of Trinil, a molar tooth was found and a month later a fossil skullcap was found. DuBois originally thought these were chimp fossils. A year later a femur (thigh bone) was found 15 meters (about 50 feet) from where the skull cap was found. The analysis of the finds showed the femur to be from an upright walking modern human and the skull cap appeared to be ape-like. Thus he concluded that the two went together and was ape-man. He named it *Pithecanthropus erectus* (upright ape-man) and concluded that it was a precursor of humans. He exhibited the find in 1895. What he didn't say was that two fossilized human skulls were found at Wadjak (on the east side of the island). He also found several more human femurs.[50]

One of the significant controversies that persists from these events is the possible failure on DuBois' part of reporting the human fossil skulls, which could account for the presence of the human femurs and reveal that bones from two different organisms were possibly combined together to form the ape-human intermediate.

Despite this relationship controversy between the skullcap and the human femurs, there is still the claim that the skullcap may have had a cranial capacity approaching 900cc. This could put these fossils in a

---

[50] Pitman, pp. 89-90.

category that is now classified as *Homo erectus*, which we will discuss later in this chapter.

Piltdown Man

The next attempt to provide the missing link was by Charles Dawson, a Sussex England lawyer and collector of fossils. For a couple of years, workers in a gravel pit near Piltdown, England would bring Dawson skull fragments which they had found. In 1908, Dawson claimed to have discovered human-like skull fragments. In 1911, he claimed to have found more. In 1912, Dawson gave his collection to Smith Woodard (Keeper of Geology at the British Museum). Woodard joined Dawson and a Jesuit paleontologist Pierre Teilhard de Chardin to conduct further work at the site. He found more bits of skull fragments and eventually found part of a right lower ape-like jawbone minus the canine teeth. The jaw was broken in two places, at the hinge points and at the point. This made it difficult to associate with the skull fragments. Also, with the absence of canine teeth it couldn't be determined if it were ape or human (apes have large canines and humans have small ones). Still, Dawson pronounced it an ancestor of the modern humans and gave it the name *Eoanthropus dawsoni* (Dawson's Dawn Man).[51]

A couple of years later, in 1913, Teilhard, claimed to have found the missing lower canine tooth in the gravel pit and the three items were assembled as ape-man. Much debate ensued over the next several years; however, it wasn't until 1953 that it was found that the whole thing was a deliberate fraud. The bones were stained to give the impression of old age and the tooth was filed to present the ape-like shape. This is an interesting story illustrating the enthusiasm for finding a missing link.[52]

---

[51] Ian Tattersall, *The Fossil Trail* (New York: Oxford University Press, 1995).
[52] Pitman, p. 92.

Peking Man

The story goes that Peking Man has its beginning as a tooth purchased at a druggist's shop in Peking, China in 1903. It was suggested that a cave at Choukoutien, China was the place to look for fossils. In 1922, two molar teeth were found there. In 1927, a third tooth was found and *Simanthropus pekinensis* was announced by Davidson Black. In 1929, Dr. Black secured a grant from Rockefeller, and established the Cenozoic Research Laboratory in Peking. Thousands of mammalian fossils and a few human teeth (575 boxes of bones) were collected through 1928. Eventually in 1929, a brain case was found to represent *Simanthropus pekinensis*, and even after Black's death in 1934, the team found more partial skulls (14 total) through the year 1937.[53]

An interesting observation was made that many of the bones were charred (and found in a seven meter deep heap of ashes) and mixed with other mammal remains including deer. The skulls had signs of being struck. Marcellin Boule was invited to the site and his assessment was that the skulls were battered monkey skulls with holes in their tops. He felt these fossils were no more important than the other animal remains. He felt that this was evidence that modern humans had been eating the mammals and much of what was found as fossils was from those meals. Also, the bones mysteriously disappeared and have never been seen since that time.[54]

Without the actual fossils, this isn't likely to be resolved. The claims are that the cranial capacity approached 900cc and therefore might be lumped in with *Homo erectus*.

---

[53] Pitman, pp. 95-99.
[54] Pitman, pp. 95-99.

Nebraska Man

Nebraska Man also had its beginning as a tooth, found in America in 1922. On the basis of this tooth, it was determined, by the American Museum of Natural History, to be a mixture of human, chimpanzee, and *Pithecanthropus*. In 1925, when the trial occurred to determine whether evolution should be taught in schools, it was used to construct a picture of what the missing link would look like. It was named *Hesperopithicus haroldcookii*; however, later it was determined to be the tooth of an extinct pig.

Recent Nominations for Human Transitional Forms

Let's turn our attention to the currently held views and nominations. Remember, the search is on to provide evidence of links between the modern human (*Homo sapiens*) and an extinct common ancestor to the modern apes. As a starting point for sorting out the nominations for missing links, let's first understand the inferred relationship between humans and other primates.

Within the order Primates, there are two suborders:

   (1) Prosimii - lemurs and tree shrews
   (2) Anthropoidea - humans, apes, and monkeys

Within suborder Anthropoidea, there are three superfamilies:

   (1) Ceboidea - New World monkeys (squirrel monkeys and spider monkeys)
   (2) Cercopithecoidea - Old World monkeys (baboons and rhesus monkeys)
   (3) Hominoidea - humans and apes

Within superfamily Hominoidea, there are three families:

(1) Hylobalidae - gibbons
(2) Pongidae - chimpanzees, gorillas, and orangutans
(3) Hominidae - humans

Evolutionists believe the prosimians were the first to diverge from an ancient ancestor, followed by New World Monkeys, Old World Monkeys, Asian Apes (orangutan and gibbon), African Apes (gorilla and chimpanzee), and then the hominids (humans). There are no intermediates to substantiate this idea of divergence of primates. One of the ways this idea is substantiated is through the concept of a "molecular clock." Like any chronographic measurement, it is based on the idea of

$$rate\ (r) \times time\ (t) = evolutionary\ distance\ (e).$$

Rate is the "known" mutation rate; *evolutionary distance* is the "determined" accumulation of those genetic mutations (based on DNA base pair similarity); and *time* is the elapsed time. Therefore $r/e = t$. The assumptions are enormous (possibly erroneous) and will not be addressed here.

Now let's look at a taxonomic comparison of apes and humans (Table 11-2), again bearing in mind that the classification is used for grouping of similar organisms, and not for inferring or conferring relationships (although this is a common mistake). There are only two ape families (Pongidae and Hylobalidae) and the human family (Hominidae) within the superfamily of Hominoidea.

Taxonomic Level	Humans	Apes	
Kingdom	Animalia		
Phylum	Chordata		
Class	Mammalia		
Order	Primate		
Suborder	Anthropoidea		
Superfamily	Hominoidea		
Family	Hominidae	Pongidae (Orangutans, gorillas, and chimpanzees)	Hylobalidae (Gibbons)

Table 11-2. Taxonomic Comparison Among Primates

The hypothetical split between the African apes and the humans is considered as the most recent of the divergences with some common ancestor. Therefore, one would think that there should be a myriad of transitional forms available to substantiate the concept. However, as we will soon see, the new nominations are few and far from convincing as missing links. What has been demonstrated, by fair paleontological study, are examples of variation (as in varieties of dog) and, in some cases, extinction.

Figure 11-2 is a high-level representation of the overall hypothesized relationship among the apes, humans, and the extinct species.

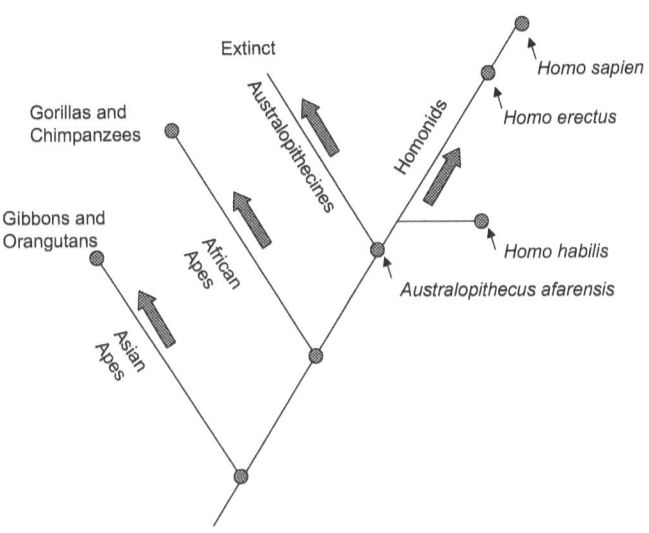

Figure 11-2. High-Level Hominoid Tree From an Evolutionary Viewpoint

A number of apelike skeletal fossils have been found in the 20th century and recently, that have been classified as a distinctive lineage of extinct species (the australopithecines) thought to have been related in this tree as well.

The structure of this hypothetical ancestral tree is the topic of constant controversy even among evolutionists. There are numerous variations appearing in the literature month to month. One species as an ancestor or predecessor of another is debated by paleontologists based on a variety of indices (*e.g.,* geographic location, morphology, geological context, assumed evolution, etc.). Even the naming and conferring of species status to various finds is possibly capricious and in the words of Tim White of UC Berkeley, "Over the past five to ten years, paleoanthropologists have gotten carried away with naming a new species every time they find a fossil with a slightly different

113

morphology."[55] The descriptions that follow provides more detail about the proposed "ancestors."

The Australopithecines

This extinct group, called *australopithecines* is classified as being members of the genus *Australopithecus* (translated - southern apemen). Furthermore, the view held by many is that specifically, one species, *Australopithecus afarensis*, was the common ancestor of the humans and other extinct species of *Australopithecus*. It is suggested by evolutionists that other australopithecine species (*i.e.*, fossils identified as *Australopithecus africanus*, *Australopithecus aethiopicus*, *Australopithecus garhi*, *Australopithecus boisei*, and *Australopithecus robustus*) are related members within Hominoidea.

Australopithecus species are characterized as averaging 1.5 meters tall and having a brain size (cranial volume) of between 300 and 500 cubic centimeters. (Note: This is very ape-like. The human cranial capacity runs in the neighborhood of 1450 to 1500 cubic centimeters).

The first australopithecine fossil was discovered and exploited in 1925 by anatomist Raymond Dart. This was a child's skull found in a South African cave (called Taung) and thus the name Taung Child. It was given the scientific classification *Australopithecus africanus*. Subsequent to that discovery, many other similar fossils have been discovered and classified as the gracile *Australopithecus africanus*, compared to other fossils of more robust types (based on a stronger masticatory system for chewing) called *Australopithecus robustus*. Robust types discovered by the Leakey's in the 1950s and 1960s at Olduvai, Tanzania have been designated *Paranthropus boisei* (or *Australopithecus boisei*).

---

[55] "Biggest *A. boisei* Cranium," *Archaeology* Newsbrief, January - February, 1998, p. 22-23.

Lucy (*Australopithecus afarensis*)

In 1974, at a site called Hadar in Ethiopia, Donald Johanson discovered a more complete set of bones classified as *Australopithecus afarensis*, (the now famous "Lucy"). This was thought to have been bipedal (walked on two legs) and closer in ancestry to humans. It had long been thought that the criteria for being a missing ape-human link was some intermediately large brain (*i.e.,* somewhere between a chimpanzee brain and the modern human sized brain). However, the brain capacity of *Australopithecus afarensis* was very small and chimpanzee-like. The conclusion that this could be a missing link of sorts — a transitional form that bridged the early apes with *Homo* — early man, was based on the possibility that the 3.5 feet tall Lucy walked on two legs instead of four. Even among evolutionists, there is a debate on the significance of these finds. There are several scientific criticisms of the find and specifically the interpretation of the joints and bone structures. These criticisms, if correct, disqualify Lucy as being bipedal and relegate her to ape status.

It can't be concluded that there is an intermediate or missing link to be found among these australopithecines or *Australopithecus afarensis.* Jean-Jacques Hublin, head of the Dynamics of Human Evolution Laboratory at France's Centre National de la Recherche Scientifique states:[56]

Contrary to a popular view, Lucy and her kin were not scaled-down models of ourselves running across the savannas. In many aspects, including social organization, australopithecines were probably much closer to apes than to humans. They were small-brained, primarily vegetarian, and preserved some ability to climb in trees.

---

[56] Jean-Jacques Hublin, "The Quest for Adam," *Archaeology*, January-February 1999, p. 30.

While Dart imagined them as early tool and weapon makers, it is still debated whether stone artifacts can be securely assigned to any of their species. Their remains in South African caves were carried there by predators, suggesting they were more often prey than hunter.

The Other Hominids

Some paleontologists, such as the Leakey's may not subscribe to the idea of a common ancestral lineage between the australopithecines and humans, but rather subscribe to the idea humans had a lineage all their own. Unfortunately, the search for a missing link in this separate lineage has produced little if any compelling evidence of transitional forms either. In a previous section we saw that Neanderthal was a fully human. A contemporary of Neanderthal was Cro-Magnon that was also undisputedly a modern human. This really leaves only two nominations to discuss: *Homo erectus* and *Homo habilis*. These so-called species lie on what may be thought of as the lineage of transitional forms from ape to human.

Homo habilis

Homo habilis appears to have a small skeleton and small brain (in the 600-800 cc range). This places them pretty far in the ape direction. The first discovery of this "species" was by Louis Leaky in 1960 at Olduvai and at the same levels where the robust *Paranthropus boisei* was found. Very early on, this fossil was rejected as being in a human lineage by paleontological scholars. In 1972, Richard Leaky (the son of Mary and Louis Leaky) discovered a skull at Lake Turkana, which had a larger brain size (775cc), and proposed its acceptance as *Homo habilis* — in the human lineage. However, this find has also been relegated to australopithecine status by Alan Walker, who was the co-author of the original paper describing the Lake Turkana specimens.

Homo erectus

Another nomination has a range of characteristics that have sort of been lumped into this species called *Homo erectus*. Some have included Neanderthal in this species. However, with the discovery of the Nariokotome boy by Richard Leaky and Alan Walker in the 1980s near the Nariokotome River (near Lake Turkana), these characteristics came to light. This was a complete skeleton, which measured about 5 feet 6 inches tall (with room to grow to an estimated 6 feet tall). Unfortunately for the evolutionists, there was no transitional morphology here.

So-called Java Man and Peking Man (if real) may qualify as this species. However, this doesn't qualify *Homo erectus* as a transitional form. The skeletal characteristics, probable bipedal locomotion, and cranial capacity approaching 900cc might just as well make them candidates as modern humans (perhaps a variety with a slightly different morphology.

Shifts in Gene Frequency

Once again, the scriptures are very clear that each species (kind) was created, not left to chance, for its form. The evolution model relies heavily on a phenomenon in nature called *natural selection*. This was a pillar of Darwinian evolution and served to further Darwin's position tremendously. Natural selection, as the name implies, is selection out of a population of plants or animals, the most fit for that environment. The interesting thing is that natural selection is a real scientific phenomenon. While most scientists would discuss it as the mechanism of evolution (and call it microevolution), it stands on its own in the creation model and in biology as a means for survival of animals in different environments. As we have already noted, variations do occur within species (*i.e.,* variation of a characteristic or variety of alleles, genetically speaking). According to the creation model, the variety of alleles for each trait such as hair color, for the most part, were created. Mutations do occur in all organisms, which in the vast majority of cases result in a fatal condition.

Natural selection is not in total disagreement with either the Bible or science. Natural selection doesn't give rise to new species, but in a Biblical model, merely allows for a shift in the gene frequency in a population to favor certain traits that give its members a better survival opportunity. Sometimes the changes in the environment are extreme and place a tremendous amount of pressure on the population. When this pressure is so extreme, it may lead to a species extinction.

Conclusion

If all of this seems confusing, it's because it is. Confusion starts when attempts are made to build a case for classifying extinct species into taxonomic categories based on fossil finds. Fragmentary evidence, uncertainty of the relationships among the bone fragments, and assumptions about the relative time periods when the fossils lived all lead to wild inferences. Furthermore, problems arise when inferences are made about the relationship of one fossil species and another contemporary fossil species. Finally, in the midst of all this uncertainty, the evolutionist suggests that there is "evidence" that these fossils can be "linked" as ancestors to modern humans.

Let's review the diversity of life predictions outlined for the evolution model:

- Transitional forms that represent the intermediates will be found as living examples.
- Transitional forms that represent the intermediates will be found as fossilized examples.
- Transitional forms that represent the intermediates will be found as fossilized examples in localized geographic areas.

It can be concluded that there is no indisputable evidence supporting any of these claims despite the relentless endeavors to produce the intermediates. Transitional forms are in a sense, the hallmark of the evolution theory. Even the staunch proponents of evolution acknowledged this from the very beginning. Yet no transitional forms have been found. This lack of transitional forms in the fossil record is undisputed among paleontologists. There has been a fervent attempt since Darwin's time to unearth a single fossil that demonstrates a transition from one species to another. Figure 11-3 is a map showing the approximate locations where the so-called missing links have been discovered.

Let's review the diversity of life predictions outlined for the creation model:

- Shifts in gene frequency will be observed in populations of species exposed to changes in environment.
- Transitional forms that represent the intermediates will NOT be found as living examples.
- Transitional forms that represent the intermediates will NOT be found as fossilized examples.

It appears that all three predictions have held for the creation model providing added confidence to the idea of distinct created kinds. The vast amounts of fossilized remains represents a history of organisms that are now extinct, however, this is to be expected given the changes that occurred after the flood. Also, in this century, we see hundreds of species becoming extinct. What the fossil record doesn't show us is any so-called transitional forms. There are no intermediates to suggest there was a transformation from one species to another — not one.

The fossil record contains examples of the very species we see today. There have been about 100,000 fossil species identified to date. These represent the fossilized remains of organisms that merely lived and died. As we have said, fossils aren't exclusively extinct species. Fossils are what remains of anything that has died subject to conditions that allowed fossilization instead of decay or consumption by another organism. In fact, many fossils are classified as belonging to modern groups (phylum, class, family, etc.) Denton indicates that, "It is still, as it was in Darwin's day, overwhelmingly true that the first representatives of all the major classes of organisms known to biology are already highly characteristic of their class when they make their initial appearance in the fossil record."[57] In other words, organisms haven't evolved.

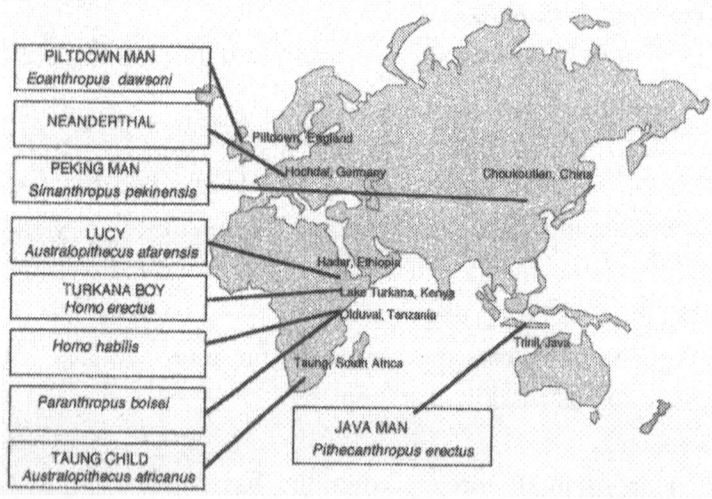

Figure 11-3. Locations of the So-called "Missing Links."

[57] Michael Denton, *Evolution: A Theory in Crisis* (Bethesda, Maryland: Adler and Adler, 1986), p. 162

CHAPTER TWELVE

The Age of the Earth

The suggested age of the earth began to get older at the same time that Darwin's theory of evolution was being proposed. The theory itself assumed a large amount of time would be available for the changes to take place. The long time needed for evolution was provided, in part, by a man named Charles Lyell in his book *Principles of Geology*, which was published in 1830. In this book, Lyell proposed the theory of *gradualism* — that the earth's features, such as canyons, river beds, mountains, and oceans can be explained by gradual weather processes (e.g., temperature fluctualtion, wind, rain, etc.) that occur today. Lyell reasoned that the earth would have to be millions of years old so that there would be enough time for various erosional and deposition processes to form the earth's geological features.

Today, geologic time is measured in two ways. One way, called *relative time,* is based on the sequential layering of the rocks forming the earth. The second way is *radiometric measurement.* In this chapter, we will examine the two methods of measuring geologic time and evaluate the predictions for the age of the earth variables. Let's begin with a review of the Age of the Earth variable and the predictions in Table 12-1.

	Model	Variable	Predictions
Age of the Earth	Evolution	The earth is somewhere around 4.5 billion years old.	- Various chronographic measurements should support an old earth.
	Creation	The earth is relatively young (less than 10,000 years.	- Various chronographic measurements should support a young earth.

Table 12-1. Predictions for the Age of the Earth Variable

Relative Time

The principle basis for relative time measurement is based on stratigraphy or layering of rock layers upon the earth (called beds) which are vertically consistent within large geographic areas and even worldwide. In other words, the sandstone, shale, and limestone layers are found in roughly the same order and thickness wherever you look. These layers, as Lyell proposed, represent geologic time. The deeper material is older than the material nearer the surface. Furthermore, with relative time, there is an added index — the fossils. This is based on the belief that evolution occurred and that less complex organisms preceded more complex ones. Therefore, when a certain type of fossil mammal is found, it is inferred that rocks in that strata (or layer) are younger since mammals were hypothetically evolved more recently. This idea becomes even more "conclusive" when the extinct forms are found, since they presumably mark a definite time frame for first appearing and then disappearing.

From a creationist's perspective, there should be no argument concerning the relative nature of the time implied by the layers of the earth. It is the geologic age that the evolutionists' assign to these layers that should cause the concern. For example, if someone stacks old newspapers in their garage, it is reasonable to consider that the papers on the bottom to be the oldest papers, and the ones nearer the top to be more recent papers. The problem with inferring the actual age of the layers are numerous and will be discussed in the next section.

Gradualism

The logic behind the evolutionists' assignment of millions of years to lower layers of the earth is based on the concept of gradualism and is one of correlation. For example, let's assume you know that water runs from your bathtub spout at one gallon per minute (gradualism). If you enter the bathroom after the tub is partially filled, and measure the amount of water in the tub and find that it contains 10 gallons, you can easily calculate that the water ran for 10 minutes. Again, the assumption is that the water runs at a rate of one gallon per minute. However, your assumption could be wrong. Let's say that whoever was filling the tub was in a hurry and decided to add buckets of water from another source or perhaps the rate of flow fluctuated. The tub could have been filled a lot quicker. Likewise under Lyell's uniform conditions of wind and rainfall, the geologic formations of the earth can be explained as having taken a very long time. However, like filling the bathtub with buckets, a catastrophic flood could get the same results much quicker.

The reason that the age of the earth is so important is that even if evolution were possible, it would require a very long time. There is little dispute that evolution has not occured in nature in the last 5,000 years. Evolutionists subscribe to an earth that is somewhere between 4.5 and 5 billion years old. However, there are a total of 68 estimates, based on observable scientific measures that fall short of the required time for evolution.[58] There are also at least 15 scientific measures that provide solid evidence of an earth that is less than 10,000 years old.[59] For example, the rate at which nickel, silicon, and lead flow into the oceans coupled with the known quantities of these elements already in the ocean, provide evidence for an earth that is less than 10,000 years. In other words, if the earth were as old as evolutionists would like to think, there should be much more of these elements found in the oceans. Unfortunately, these very young earth measures are all based on the same assumptions of uniformitarianism (*i.e.,* the rates of decay or disintegration, or erosion are occurring at the same rate today as

---

[58] Henry M. Morris and Gary E. Parker, *What is Creation Science*, (El Cajon: Master Books, 1987), pp. 285-293.
[59] Morris and Parker, pp. 285-293.

they were thousands of years ago) and gradualism. The concept of gradualism also led to the concept that the deeper you go in sedimentary rock, the older the fossils are that were found in that strata (Table 12-2). It is important to note here that this geologic time scale was largely determined well before any of the radiometric methods were developed or employed. The point is that, like many other evolution concepts, a predetermined conclusion was made.

Gradualism is also the basis for explaining the evolution of organisms of lower complexity (found at lower levels) to organisms of higher complexity (found at higher levels). For example, trilobites (Figure 12-1) are found at the bottom of the time scale in the very early Cambrian period.

Era	Period	Epoch	Start Time	Life Forms
CENOZOIC	QUARTERNARY	RECENT	10,000 yrs ago	
		PLEISTOCENE	2,500,000	Man
	TERTIARY	PLIOCENE	12,000,00	Grazing and
		MIOCENE	26,000,000	Carnivorous
		OLIGOCENE	38,000,000	Mammals
		EOCENE	54,000,000	
		PALEOCENE	65,000,000	
MESOZOIC	CRETACEOUS		136,000,000	Primates and Flowering Plants
	JURASSIC		195,000,000	Birds
	TRIASSIC		225,000,000	Dinosaurs and Mammals
PALEOZOIC	PERMIAN		280,000,000	
	CARBONIFEROUS	PENNSYLVANIAN	320,000,000	Reptiles and
		MISSISSIPPIAN	345,000,000	Fern Forests
	DEVONIAN		395,000,000	Amphibians and Insects
	SILURIAN		430,000,000	Vascular Land Plants
	ORDOVICIAN		500,000,000	Fish and Chordates
	CAMBRIAN		570,000,000	Shellfish and Trilobites
PRE-CAMBRIAN			4,650,000,000+	Algae, Eucaryotic Cells, and Procaryotic Cells

Table 12-2. The Concept of Hypothetical Geological Time Scale.[60]

---

[60] "Geology," *Funk and Wagnals New Encyclopedia*, 1986 ed.

125

Figure 12-1. A fossil trilobite three inches long.

The evolutionists' explanation is that the trilobite was an early life form, thus found in lower strata. However, consider that trilobites were exclusively ocean dwelling and lived on the ocean bottom. A better explanation might be that a catastrophic flood (*i.e.,* the Genesis Flood) and its associated violent activities would have silted over the lowest dwelling organisms first. The trilobites would have been prime candidates for a lower strata. If an organism's ecological niche and its capabilities for relocating (*e.g.,* climbing a tree, or flying) are considered together, it makes sense that so-called higher organisms are found in higher strata. As we saw in the creation model, the various strata would contain the fossilized remains of life destroyed in the Flood.

The creation model presents a strong case for catastrophism. Gish and others point out that the age of the earth can never scientifically be proven. What can be said is that there is a significant variation in the results of the so-called geologic chronometers (or clocks). Therefore, the evolutionists can't be absolutely sure about their age of the earth. The 4.5 billion year age of the earth cannot be supported by any accurate means of measurement. This quote from the Plummer and McGeary *Physical Geology* textbook sums up the weak arguments:

> The 4.5 billion-year estimated age of the earth comes from various lines of evidence, mostly worked out by astronomers and planetary geologists who feel that all the planets formed about this long ago.[61]

This type of conclusion is not science and the 4.5 billion year age of the earth cannot be supported with words like "estimated," and "mostly worked out," and "feel," and "about."

Radiometric Time

The fact of radiometric decay was discovered by a French physicist names Henry Becqueret in 1896. The premises for radiometric dating are quite simple; however, the assumptions and techniques raise questions concerning the accuracy and meaning of the results.

Radiometric Dating

There are several radiometric dating methods in current practice. Each has its own set of limitations and applications, which we will not discuss here. There are two main radiometric methods we will discuss: (1) Uranium - Lead and (2) Carbon-14. The Uranium - Lead method is used for dating inorganic material (rocks) and is illustrative of the other rock dating methods (e.g., Potassium - Argon). Carbon-14

---

[61] Charles C. Plummer and David McGeary, *Physical Geology* (Dubuque: W. C. Brown Company, 1982), p. 154.

is used for dating material that was once living and still contains carbon from its original living composition.

Before discussing the mechanism of radiometric dating, let's review the definition of radioactivity. Atomic nuclei of radioactive elements, such as uranium-238 (92 protons and 146 neutrons), are unstable and break down into smaller atoms. For example, over time, uranium-238 loses 10 protons and 22 neutrons, leaving it with only 206 nuclear particles which is now nonradioactive lead (Pb - 206). This is referred to as a daughter product. The rate of so-called decay has been measured and is known for these radioactive elements. The time it takes for 50% of the element to decay completely (*e.g.,* the parent isotope of uranium to go to the daughter product lead; or carbon-14 to go to nitrogen-14, is called its half-life).

The calculations can be thought of in terms of the following formula: [62]

$$ t = \frac{1}{\lambda} \ln \left(1 + \frac{D}{P} \right) $$

t is time (or the calculated age of the sample).
λ is the rate of decay.
D is the amount of daughter product in the sample.
P is the amount of the parent isotope in the sample.

The assumptions necessary to calculate the unknown (t) are:

1   The decay rate (λ) is known.
2   The decay rate (λ) has always been constant.

---

[62] William L. Newman, "Radiometric Time Scale," *Geologic Time*, (U.S. Geologic Survey) http://pubs.usgs.gov/gip/geotime, July 1997.

3   The amount of the daughter product in the sample at the time of its formation is known (or is 0).

4   There has been no parent isotope or daughter product introduced or removed from the sample by any means (other than by radioactive decay) since the sample's original formation.

Uranium to Lead

In this method, scientists assume that the half-life for U-238 is 4.5 billion years. This means that if you had 10 grams of U-238, after 4.5 billion years you would have only 5 grams of this parent isotope and 5 grams of lead. Let's critically consider each of the assumptions listed above:

1   Scientists know what the current rate of decay is for radioactive isotopes.

2   It is a major assumption that the decay rate has remained constant, but there is no proof that it has. It appears that uranium-238 decay is constant based on observations since the decay concept was discovered. However, whether or not this assumption holds up over the 4.5 billion years can't be definitively answered. For example, it is known that the decay rate for Thorium-234, is slower (half as fast) 25 days after being isolated from uranium.[63]

3   The basis for knowing the amount of daughter product in the original formation is a tricky one because the answer is a guess. If all the uranium-238 was formed at some supernova event (that also formed the earth) much of the decay of U-238 to lead may have already occurred. Therefore to date the earth based on a beginning ratio of uranium to lead is incredibly presumptuous. Even if the other assumptions are controlled,

---

[63] "Radioactivity," *Funk and Wagnals New Encyclopedia*, 1986 ed.

the material being dated may reflect the age of the supernova event or the age of the universe — not the age of the earth.

4   Uranium and other radioactive elements can be lost from the rock by dissolving in weak acid washing over it through time. There are numerous problems associated with this assumption. Sedimentary rocks (such as ones formed from a flood) can't generally be dated because of the possibility they could have been contaminated by previous existing or previously formed daughter products. Other problems include the possibility of loss of material during weathering and heat.

Carbon-14

Radioactive carbon-14 (C-14) is formed in the upper atmosphere from cosmic rays and nitrogen gas. Non-radioactive carbon-12 (C-12) is also present. Both of these carbons are found in carbon dioxide gas and are taken in by plants for growth. Animals eat plants and thus consume C-12 and also C-14. C-14 analysis actually dates the age of death of the organism; when the organism stops taking in C-14 and C-12. C-14 decays away and C-12 does not; so the method measures the ratios expected based on the decay rate. Less C-14 means an older specimen and more C-14 means a younger specimen. The half-life of C-14 is about 5,730 years. Since C-14 is made in the atmosphere, one assumption is that its availability has been constant over time, but that is not the case. It has been shown that volcanic activity causes variation in the amount of C-14 present. Also, in 1918 a comet fell over Siberia and the amount of C-14 in the atmosphere doubled. Solar flare activity and ozone pollution are two other factors shown to affect C-14 levels in the atmosphere. In pre-flood days when there was a vapor canopy that prevented entry of cosmic rays then there would have been little or no C -14. Preserved organic material from pre-flood times then would be dated as very old when actually they would be a few thousand years old.

Conclusion

There is no substantial, irrefutable evidence that the earth is 4.5 billion years old. Unfortunately, the chronographic measures that might show the earth to be very young (*i.e.,* less than 10,000 years old) rely on the same uniformitarian concepts. Therefore it is difficult to say with certainty, exactly how old the earth is. There are tremendous problems associated with accepting the assumptions for radiometric dating methods — at least enough to justify questioning the validity of the claims concerning the ages associated with fossil and geologic specimens.

CHAPTER THIRTEEN

Final Thoughts

The dichotomy between evolution and creation is not one of science versus non-science. As we have seen, the evolution premises and many of the findings themselves have raised serious questions about the methods and motives of the researchers. The evolutionist has nothing to be proud of in this regard. On the other hand, it appears that some creationists can become very close minded regarding true science. The "God said it — and that settles it" attitude may reflect right thinking, but turns a blind eye to some of the great mysteries of nature. Einstein said, "Science without religion is lame, religion without science is blind."[64] Although not all creationists are Christians, there are some guidelines we might consider from the Apostle Peter. In his first letter he writes "Always be prepared to give an answer to everyone who asks you to give the reason for the hope you have. But do this with gentleness and respect." (I Peter 3:15)

Dogma

Dogma is essentially a principle that is accepted as an absolute truth. It may have roots in well tested theory, or possibly just blindly accepted. Things that have been historically accepted as true aren't often challenged and retested. This can take a scientist pretty far down a wrong path, and eventually lost in the woods. Here are some examples of dogma that serve to alert us to the possibility of other questionable absolutes.

---

[64] Abraham Pais, *Subtle is the Lord. The Science and the Life of Albert Einstein,* (Oxford: Oxford University Press, 1982), p. 319.

Kettlewell's Peppered Moths and Natural Selection Dogma

A very common example of natural selection used in biology textbooks are the peppered moths, *Biston bitularia,* in 19th century England. There exists two alleles for wing color; dark peppered and a light peppered color. Light peppered moths predominated in pre-industrial England. Hypothetically, since there was no black soot on the trees, light peppered moths blended in on the bark and were not eaten by birds, while the few dark moths were eaten more frequently. Then after the industrial revolution, the tree bark became black from soot and the light peppered moths were eaten, but dark moths survived and increased in numbers. Thus the number of black moths became greater than the number of light peppered moths. This was considered by many creationists as innocuous since it represented a shift in the gene frequency of the dark and light peppered color alleles (more dark ones and fewer light ones). If conditions changed to favor the light colored moths, the ratio would change to more light ones and fewer dark.

In the 1950s the British scientist, H. B. Kettlewell conducted experiments that supposedly proved the moths were distributed differently based on the environmental pressures and the likelihood of predation was increased for the non-adapted moths. Kettlewell also suggested that when he released his experimental moths, they preferred to settle on backgrounds that matched their wing coloration (e.g. dark moths to dark trees). There have arisen a number of criticisms and problems associated with Kettlewell's experiments and methods. First, he released the moths by placing them directly on the tree trunks. Second, it turns out that the *Biston bitularia* moth doesn't rest on tree trunks. Third, the behavioral experiments indicating *Biston bitularia's* preference for backgrounds matching their coloration have been unsubstantiated despite repeated efforts to duplicate the findings.

Despite this overturning of what was once considered proof of natural selection as a force in evolution, the story lives on in the textbooks.

Haeckel's Ontogeny Recapitulates Phylogeny Dogma

We can remember back in undergraduate biology when we learned a phrase which almost became a mantra chanted by the biology student. It was "ontogeny recapitulates phylogeny." Ontogeny refers to the stages of embryonic development organisms go through from the fertilization of an egg onward. Phylogeny is the inferred or conferred stages of evolution that hypothetically occurred in that organism's evolutionary history. For example, in a human, the egg stage is analogous to a single cell protozoa. The human embryo had stages analogous to a frog, then a fish, etc. This idea was proposed by a contemporary of Darwin named Ernst Haeckel in the mid 1800s. Soon after Haeckel's ideas, many scientists criticized the recapitulation theory and even suggested that Haeckel may have falsified his illustrations.[65] Unfortunately, the concept has somehow persisted to this day in some classrooms. In a recent letter to the journal, *Science*, the authors of a new evaluation of Haeckel's work wrote, "Unfortunately, Haeckel was overzealous. When we compared his drawings with real embryos, we found that he showed many details incorrectly. He did not show significant differences between species, even though his theories allowed for embryonic variation."[66]

Where's the science in that? This wasn't a small "oops I missed a comma." This was terrible science and the result was an evolution proposition that became dogma — an absolute. The conclusion was Haeckel was overzealous?

---

[65] Wayne Frair, "Embryology and Evolution," *Creation Research Society Quarterly,* Vol. 36, No. 2, September 1999, pp. 62-67.
[66] Michael K. Richardson *et al*, "Haeckel, Embryos, and Evolution," *Science,* Vol. 280, 15 May 1998, p. 983.

Nerve Cell Non-Regeneration Dogma

This is actually good news. Although a little off the evolution track, here is an example that makes the point that even the well-meaning, learned, and enlightened modern scientists can, and do, discover that — yes, they can be wrong. We chose this example to show the positive effect an absolute change of thinking can have on the world of science and medicine.

Up until recently, it had been accepted that the human brain, when damaged, doesn't repair its neural tissue. The reason was thought to be because it lacked the undifferentiated cells (stem cells) that would facilitate the necessary regeneration. A discovery by Peter Erickson et al, provides evidence that the human brain can and does "spawn neurons routinely in at least one site - the hippocampus, an area important to memory and learning."[67] This is a commendable example of not accepting what was thought to be dogma. The lead-in to the article says, "Contrary to dogma, the human brain does produce new nerve cells in adulthood."[68]

Will the Emperor Ever Acknowledge He is Naked?

The story of the emperor and his new suit is a fascinating story. Its moral and application is obviously quite broad and the metaphor plentiful. The emperor hoped he would have splendid clothes and he hoped others (the worthy ones) would be able to see and appreciate them. This is not the kind of *hope* we referred to in the quote by the Apostle Peter. The hope that is reflected by people like the emperor and his courtiers is wishful thinking at best. Even in the Hans Christian Andersen story, it is obvious that all hope (even in that

---

[67] Gerd Kempermann and Fred H. Gage, "New Nerve Cells for the Adult Brain," *Scientific American,* May 1999, p. 48.
[68] Kempermann and Gage, p. 48.

sense) was gone. So why hang on to a unsubstantiated claim? Perhaps there is a clue in the lyrics to an old Simon and Garfunkel song that says "Still, a man hears what he wants to hear and disregards the rest."[69]

In a more concrete example, let's consider a story, from an evolutionist's viewpoint, that illustrates what we call *thought inertia*. In 1994 a short article appeared in the British medical journal *Lancet*, entitled "Darwin, have I failed you?" R.V. Short, a medical school professor at Monash University Australia wrote of an experience he had with his medical school class. Professor Short wanted to introduce his students to some origin of man history, as a way of getting them to think about some larger social issues. He decided to issue a questionnaire to his approximately 150 students to determine their belief regarding evolution. He found that 27% thought Darwin was wrong; 27% did not believe in an ape-like ancestor for humans; and 21% believed that Eve was created by God from Adam's rib. Professor Short went to some extremes to try to enlighten the 27% into the realm of evolutionary thought. He designed a course for all the students consisting of eight one-hour lectures on origin of species, evolution, apes and the emergence of humans, etc. He showed films and even visited the Melbourne Zoo. This was all supplemented by a variety of required reading on evolution topics. At the end of the class, the students were surprised by getting the original questionnaire to answer. Short was shocked at the outcome. "To my utter dismay, there were no statistically significant changes in any of the answers to any of the questions. I was shattered. I believe in the truth of evolution and still regard it as the most exciting fundamental concept that underpins the whole of biological thinking."[70]

Two interesting points can be made from this story. First, people aren't going to change their minds about something they believe

---

[69] Paul Simon, "The Boxer," *Bridge Over Troubled Water,* 1970.
[70] R.V. Short, "Darwin, Have I Failed You," *Lancet*, Vol. 343, 26 Feb 1994, pp. 528-529.

strongly about. In this case, apparently the 27% that rejected evolution didn't move from their position. However, it would have been interesting to try a similarly structured eight week course on creation topics then see if that had an impact on the 73% that had originally accepted evolution. The second point is perhaps the more illustrative. Notice Professor Short's defense of his own beliefs about evolution. Why did he seem compelled to restate his acceptance of the dogma of evolution? One can only speculate, but it is worth noting that just like in the Emperor's New Clothes, one child speaking out about the truth began the all out cry that the emperor was naked.

Many scientists that are creationists believe that we are at a turning point. Some believe that within a few years, support for evolution will rapidly decay (unlike the radioactive decay of U-238).

Keep an Open Mind

Real science keeps an open mind. Creationists are often characterized as not having an open mind when it comes to origin and diversity of life issues. It is true that we tend to place certain boundaries (*i.e.,* Biblical constraints) around the possible solution space. However, this is still scientific and prudent. This is the basis for the scientific hypothesis, otherwise there is no basis for testing. For example, in mathematics, constraints and variables are used to define the boundaries within which the solutions may lie.

If it weren't for boundaries, we would be constantly chasing after implausible ideas such as flying trees. If we hypothesized that trees once flew, what would keep me from securing research resources to study that idea? There might be empirical evidence that they once flew, such as their shape, their leaves' affinity for windy conditions, or their appearance at a wide variety of locations geographically. Will we ever know if trees flew? Should we design experiments to test plausible hypotheses? The answer is obviously, no. Why? Because we know enough about the tree, its morphology and its place in the biological world. Flying trees, although simple-minded, is an example of the kinds of things that would fall outside of the boundary of possible solutions. In the same way, we can agree to examine diversity of life, but within a boundary of Biblical reasonableness.

As we pointed out in an earlier chapter, there are thousands of creation scientists that are working, researching, practicing, and teaching in the highest positions. Evolutionists seem to want to characterize the creationist as a pseudo-scientist, but nothing could be farther from the truth.

On the Horizon

It is difficult to say with any certainty what lies ahead with regard to the many elements that comprise the evolution creation controversy. The battle is fought on many fronts (*e.g.,* education, research, politics, religion, law). It is not a battle that is apt to be won or lost by persuasive argument or by cleverness of speech. The person who genuinely seeks to embrace the truth will find it. Support for evolution is declining and the scientific support for creation is increasing. If this is any indication of the direction we are headed, creation and intelligent design theory should be popular soon. The foundations of evolution continue to receive criticism and are somewhat shaken. On the other hand, with every turn of the archaeologist's shovel, there seems to be growing support for the veracity of Holy Scripture.

The Genesis Flood

We already pointed out that the catastrophic flood of Noah (Genesis Flood) is a critical component of the creation model. It impacts the course of civilization, the earth's flora and fauna, the geologic structure of the earth, the weather, the ice age, and even some of the fossils. An interesting headline appeared in the Washington Post last year. It read, "Black Sea Artifacts May Be Evidence of Biblical Flood."[71] The article reported the now-famous discoveries from Robert Ballard (who led the Titanic discovery) and his National Geographic Society supported team. "The team discovered the outlines of an ancient coast 550 feet below the current waterline, the first visual evidence that a flood had occurred in the region eons ago."[72]

Conclusion

The question remains: Does it matter where we came from? The answer is yes. However, merely agreeing that it matters may not alter one's course or direction. We have examined the history of evolution and the foundations upon which it is built. Without question, there seems to be an unrelenting zeal for justifying a concept that dismisses God and replaces Him with a flimsy, nonretributive alternative. Therein lies what seems to be the crowning glory of evolution — the abandonment of the need for embracing the divine sovereign Creator. Without God, the world would seem to be without inherent accountability. In the first part of the book, we focused on people as the centerpiece of the creation. Undeniably, people have some innate need for seeking their roots and having some idea of where they are going. God has revealed the answer to both questions in a couple of very special ways. One way is through the creation itself.

Anyone, whether or not he or she is a scholar, is aware of the world around them. For the scientist (even a science student) the complexities of the universe are enormous and wonderful. The vastness of the macrocosm of space and its mystery and beauty is

---

[71] Guy Gugliotta, "Black Sea Artifacts May Be Evidence of Biblical Flood," *Washington Post* (Online), 13 September, 2000, p. A1.
[72] Gugliotta, p. A1.

excruciatingly complex and fits together with perfection. The microcosmic world is likewise beautiful and complex. Yet no person can even come close to creating what has already been created. We can only observe and analyze and try to comprehend it.

Appendices

APPENDIX A

Science and the Scientific Method

Prior to 1859 and the evolution theory, science was not viewed as antagonistic to Christianity. Science was actually a friend of religion and was viewed as providing continuing support for the idea of wisdom of a creator and the grandeur of His design. The introduction to the first issue of the prestigious *Zoological Journal of London*, published in 1824, made this statement: "The naturalist...sees the beautiful connection that subsists throughout the whole scheme of animated nature. He feels too that at the head of all this system of order and beauty, preeminent in the domain of his reason, stands Man...the favoured creature of his Creator."[73]

In 1857, one of the leading biologists of North America, Louis Agassiz, Professor of Zoology at Harvard, wrote that the living world "shows also premeditation, wisdom, greatness, prescience, omniscience, providence...all these facts proclaim aloud the One God whom man may know".[74]

In this appendix, we will look at what science is and what makes up the scientific method. This will form a foundation which will make the concepts of the creation model more understandable. This foundation will provide a basis for critically examining the evolution model as well.

Science

Science is used to study the laws of the physical (non-living) and biological (living) world. These are the two main divisions of pure

---

[73] Michael Denton, *Evolution: A Theory in Crisis* (Bethesda, Maryland: Adler and Adler, 1986), p.20
[74] Denton, p. 20.

natural science. Within each of these divisions are disciplines or fields. Some disciplines within the biological sciences include:

Discipline	Area of Study
Botany	Plants
Zoology	Animals
Ecology	Organisms & Environment

The physical sciences are disciplines concerned with the sciences of matter and energy. Some areas of study within the physical sciences are:

Discipline	Area of Study
Chemistry	Matter
Physics	Energy and Matter
Astronomy	Planets and Stars
Geology	Rocks and Minerals

Science has illuminated many mysteries of creation. Many scientific discoveries have gone far to improve our quality of life. Every day, new discoveries are made in the disciplines of biology, medicine, and chemistry that save lives and improve our health. We owe much to the men and women who endeavor to understand our world and make it a better place to live. Many of these scientists realize that God created the universe and it is God that allows these discoveries to be made.

The Scientific Method

Scientists use the scientific method to study the biological and physical world. Personal opinions, family traditions, an explanation of a situation from friends, or cultural traditions are not part of the scientific method. There are six basic steps that comprise the scientific method.

Step 1 - Identify the Problem

Identify the problem or phenomenon to be investigated. Generally a scientist already has accumulated many facts and made observations about the problem.

Step 2 - Formulate an Hypothesis

An hypothesis is an educated guess or explanation of the problem and is often stated as an IF - THEN statement. As an example: IF red food dye is fed to mice, THEN they will develop stomach cancer. The hypothesis must be testable in an experiment. Another way to state the hypothesis is as a null hypothesis. The null hypothesis is an hypothesis that is written in the negative. For example: Red food dye does not cause stomach cancer in mice. If during experimentation, stomach cancer is found in the mice getting the red dye, then the hypothesis is proved wrong (rejected) and one can say red food dye might cause cancer in mice.

Step 3 - Test the Hypothesis

The hypothesis is tested by an experiment where data is collected and then analyzed. The experiment will involve observations of some type. Usually the observations will be measuring or counting something such as the number of mice that develop stomach cancer. These observations are known as data, which can be mathematically analyzed. Very often experiments involve two groups. One group is the control group and nothing experimental is done to this group — but it is needed to see if an unknown factor could be causing some phenomenon to occur. The other group is called the experimental group and this is the group that is experimentally tested. Using the previous example, one group of mice could be fed red food dye in their food to see if they get stomach cancer. This is the experimental group. Another group of mice is fed the same food without the red dye and is, in every way possible, treated the same as the experimental group. These mice are called the control group. It is called the control group because the scientists want to make sure it is the dye and not something else that causes the cancer. If the control group were to get cancer, then something else other than the red food dye may be the cause.

Step 4 - Analyze the Data and Formulate Conclusions

This is actually the hardest part of the scientific method. Although the data analysis is presented objectively, there may be some speculation or opinion involved in the interpretation. Usually the experiment will either support or reject the hypothesis.

Step 5 - Report the Data in a Scientific Journal

This is important so others can read and think about the experiment. The report includes the hypothesis and details of how the experiment was conducted, as well as the results.

Step 6 - Verify the Hypothesis

Other scientists should be able to duplicate the experiment and get the same results. This serves to verify that what was found in the experiment was valid.

APPENDIX B

Essential Chemistry and Biochemistry

Chemistry is the study of the composition, properties, and behavior of the material world. Chemists study how substances are formed and learn to predict under what conditions substances will react with each other. Chemistry is also concerned with the states of matter (*i.e.,* solid, liquid, or gas). The world around us is comprised of various objects. These objects are made of materials that may be pure (such as gold) or may be combinations of various substances such as plastic. Even water is a combination of two substances: hydrogen and oxygen.

In order to understand the creation and to be able to understand the claims of the evolution model, it is important to have a basic understanding of chemistry. Understanding the origin of life, the age of the earth, and organization of living things depends on an understanding of chemistry. We will consider some of the radiometric dating techniques applied by evolutionists to substantiate their claim of an old earth. Our study of chemistry will allow us to understand the methods and the shortcomings of these techniques. In this chapter we will begin by explaining atoms, molecules, and compounds; and then organic compounds found in living organisms.

Atoms and Elements

Atoms are the smallest unit of matter. One cannot split or divide an atom by normal chemical means. However, under special conditions the atom can be split releasing tremendous amounts of energy (*e.g.,* the atomic bomb or a nuclear power plant). There are over 100 different kinds of atoms which we call elements. An element is a pure substance that is made of one kind of atom. For example, oxygen atoms, hydrogen atoms, gold atoms, silver atoms, and carbon atoms are a few of the elements.

All atoms have a central nucleus with electrons surrounding it. The central nucleus contains protons and neutrons. Each proton has a positive electrical charge. Neutrons have no electrical charge. Neutrons and protons are very small but very heavy. 1/3 teaspoon of neutrons and protons from any element that are packed together would weigh 100 million tons. However, we know that 1/3 teaspoon of an element does not weigh than much. Even 1/3 teaspoon of gold, which is very heavy, doesn't weigh much. Why? The answer is because there is a lot of empty space in a teaspoon of gold. The space occurs between the nucleus and electrons.

Electrons have a negligible weight and a negative electrical charge. They are arranged in layers (called orbitals) and move around the nucleus. Electrons with the least energy are in orbitals closest to the nucleus. Electrons with the most energy are in orbitals farther away from the nucleus.

The structure of the helium atom, which has only two electrons is represented by the drawing in Figure B-1.

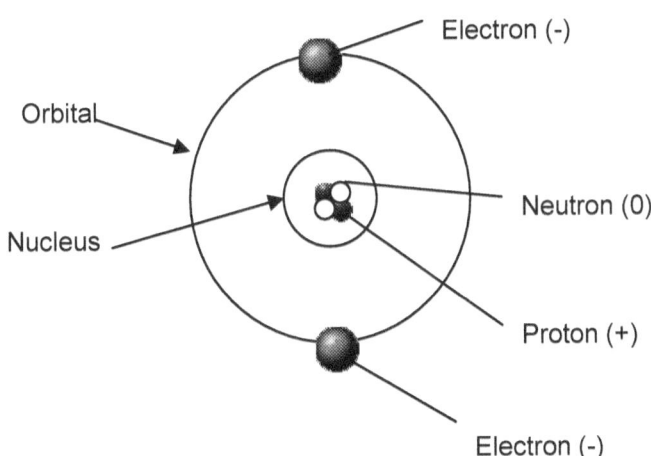

Figure B-1. A Helium Atom

Elements are often abbreviated by letters (*e.g.,* C for carbon; O for oxygen). Also, elements are numbered according to the number of protons they contain. We start with number 1 and continue to more than 100. This number is called the atomic number. The atomic number equals the number of protons. The atomic number also equals the number of electrons in a neutral atom.

Some examples:

- Hydrogen has the atomic number 1; it has 1 proton and 1 electron.
- Oxygen has the atomic number 8; it has 8 protons and 8 electrons.
- Nitrogen has the atomic number 7; it has 7 protons and 7 electrons.
- Sodium has the atomic number 11; it has 11 protons and 11 electrons.

The atomic weight equals the number of protons plus the number of neutrons. Protons and neutrons are responsible for the weight of an atom and as we said before, electrons are practically weightless. Examples of atomic weight:

- Carbon: atomic weight of 12 and atomic number is 6. Atomic wt. (12) = protons (6) + neutrons (6).
- Hydrogen: atomic weight of 1 and atomic number of 1. Atomic wt (1) = protons(1) + neutrons (0).
- Oxygen: atomic weight of 16 and atomic number of 8. Atomic wt (16) = protons (8) + neutrons (8.)
- Nitrogen: atomic weight of 14 and atomic number of 7. Atomic wt (14) = protons (7) + neutrons (7).
- Sodium: atomic weight of 23 and atomic number of 11. Atomic wt (23) = protons (11) + neutrons (12).

The following chart (Table B-1) puts it all together.

Element	Abbr	Atomic #	Atomic Wt	Protons	Neutrons	Electrons
Carbon	C	6	12	6	6	6
Hydrogen	H	1	1	1	0	1
Oxygen	O	8	16	8	8	8
Nitrogen	N	7	14	7	7	7
Uranium	U	92	238	92	146	92
Lead	Pb	82	207	82	125	82
Potassium	K	19	39	19	20	19

Table B-1. Sample Chart of Element Composition

Isotopes

Sometimes elements have a variable number of neutrons in their nucleus. When this occurs, the element is said to have isotopes. For example, carbon in its most common form has an atomic weight of 12 (6 protons and 6 neutrons) and is called carbon-12. However, there are small quantities of carbon-13 (an isotope of carbon) which has 6 protons and 7 neutrons. Another isotope occurs as carbon-14 which has 6 protons and 8 neutrons. carbon-12, carbon-13, and carbon-14 are all isotopes of carbon.

Another example is hydrogen which has three isotopes. hydrogen-1 (protium), is the most common and has one proton and no neutrons. Hydrogen-2 (also called deuterium) has one proton and one neutron. Finally, hydrogen-3 (or tritium) has one proton and two neutrons.

An isotope that decomposes (*i.e.,* the nucleus changes) spontaneously is called a radioactive isotope. Radiation is emitted by radioactive isotopes and can be dangerous or fatal to living organisms in high amounts. The decomposing isotope gains or loses protons and may also gain neutrons. As a result, the radioactive isotope becomes a different element called a daughter product. Some isotopes break down very quickly and some take many years. Two important isotopes of uranium are uranium-235 and uranium-238. Both the uranium-235 and the uranium-238 isotopes break down to lead. For example, U-238 begins with 92 protons and 146 neutrons. Through a series of complex steps over time, it loses 10 protons and 22 neutrons, leaving it with only 206 nuclear particles which is now the lead isotope (Pb-206). The time it takes for 50% of the original U-238 to become Pb-206 is called its half-life. The half-life of uranium-238 is said to be 4.5 billion years.

The application to dating rocks is based on knowing the amount of uranium and lead in a given sample. If a rock originally contained ten grams of uranium, it is presumed that after 4.5 billion years there would be five grams of uranium and five grams of the daughter product lead formed. Obviously, certain assumptions must be made. First, the relative amounts of original uranium and lead must be known. Also, it is assumed that the decay rate has not changed since the formation of the original rock. The third assumption is that nothing has occurred over time to change the amounts present for either element. These are three assumptions that bring the radiometric dating process, used by evolutionists, into question.

Compounds and Molecules

A compound is a substance that contains two or more different elements. For example, water contains the elements hydrogen and oxygen. Glucose contains hydrogen, oxygen, and carbon. Carbon dioxide contains the elements of carbon and oxygen.

A molecule contains two or more atoms. A molecule is the smallest amount of a compound that can occur. For example a molecule of water contains two atoms of hydrogen and one atom of oxygen and is written H_2O. A molecule of glucose contains six atoms of carbon, six atoms of oxygen and 12 atoms of hydrogen and is written $C_6H_{12}O_6$. Oxygen is a gas and occurs as a molecule containing two atoms of oxygen (*i.e.,* O_2).

Figure B-2 shows drawings of water, oxygen, and carbon dioxide molecules.

There are two types of molecules: those that contain the carbon atom and are called organic molecules (*e.g.,* glucose) and those that do not contain the carbon atom and are called inorganic molecules (*e.g.,* water). There are four large organic compounds that make up living things. These are proteins, carbohydrates, lipids (fats) and nucleic acids.

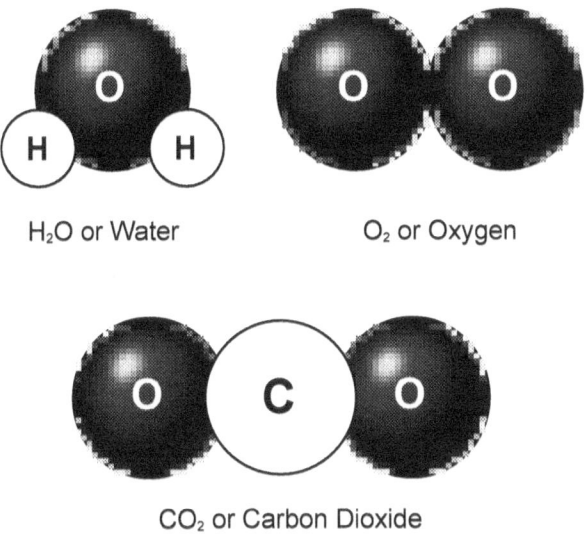

H₂O or Water   O₂ or Oxygen

CO₂ or Carbon Dioxide

Figure B-2. Molecules

Proteins

Proteins are polymers (chains) of amino acids. They are found in the skin, hair, body tissues (*e.g.,* muscle), and enzymes. Amino acids are chemically strung together (like beads on a bracelet) to form a protein. There are 20 basic amino acids found in plants and animals, and each one has an abbreviation such as Ala for alanine. The amino acid general structure is shown in Figure B-3.

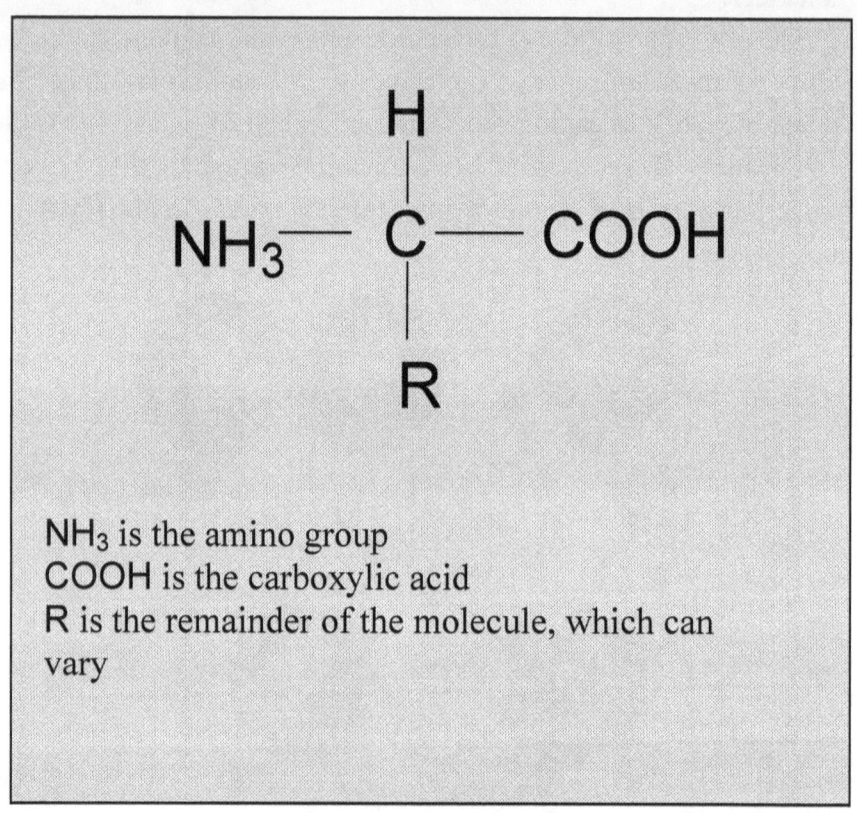

NH_3 is the amino group
COOH is the carboxylic acid
R is the remainder of the molecule, which can vary

Figure B-3. Amino Acid Structure.

The 20 different amino acids have different R groups. Table B-2 is a list of the 20 different amino acids (and their abbreviations) that make up all living things:

Name	Abbreviaton	Name	Abbreviaton
Alanine	Ala	Leucine	Leu
Arginine	Arg	Lysine	Lys
Asparagine	Asn	Methionine	Met
Aspartate	Asp	Phenylalanine	Phe
Cysteine	Cys	Proline	Pro
Glutamate	Glu	Serine	Ser
Glutamine	Gln	Threonine	Thr
Glycine	Gly	Tryptophan	Trp
Histidine	His	Tyrosine	Tyr
Isoleucine	Ile	Valine	Val

Table B-2. Amino Acids.

The primary structure of proteins is determined by the sequence of amino acids. For example, the hemoglobin protein carries oxygen in the red blood cells and is actually made up of four protein chains. Below are examples of both the correct and incorrect sequence of amino acids in one of the protein chains of hemoglobin.

Correct Sequence: val-his-leu-thr-pro-**GLU**-glu-lys...
Incorrect Sequence: val-his-leu-thr-pro-**VAL**-glu-lys...

The sequence or order of the amino acid is very important in the protein. The wrong sequence of one amino acids causes the protein to be different and non-functional. You can see that the valine amino acid is the incorrect amino acid. When this happens in the human being, the hemoglobin protein doesn't function correctly. The result is the person has sickle cell anemia and is unable to carry the correct amount of oxygen in the body.

Carbohydrates

Carbohydrates include the sugars, starches, and fiber. They are used for energy in plants and animals and also as structural components. Carbohydrates occur in three different forms. These are monosaccharides, disaccharides, or polysaccharides.

Monosaccharides are simple sugars that contain one ring-like structure of carbon atoms with oxygen and hydrogen atoms attached to the carbon atoms. The three most common monosaccharides are: (1) Glucose also called corn sugar or dextrose. (2) Fructose from fruits. (3) Galactose from milk.

Disaccharides are two simple sugars chemically bonded together. Three examples are: (1) Sucrose from plants made up of glucose and fructose. (2) Lactose from milk made up of glucose and galactose. (3) Maltose from grain seeds made up of glucose and glucose.

Polysaccharides are long chains of monosaccharides (simple sugars) chemically bonded together. Some examples are: (1) Starch is a storage form of glucose found only in plants. A starch molecule contains 3,000 to 18,000 glucose molecules; so a starch molecule is very large. (2) Glycogen is the storage form of glucose found only in animals. (3) Cellulose is structural component found in plants. Animals can't digest it.

Lipids

Lipids are water insoluble substances. We will look at just four common groups of lipids.

Fats and oils (or triglycerides). Fats are solids at room temperature and come only from animals. Some examples are butter, margarine and triglycerides which are found in your blood. Oils are liquids at room temperature and come only from plants. Some examples are corn and olive oil.

Phospholipids are another group of lipids with a similar structure to fats and oils except that they contain the phosphate group (composed of a phosphorus atom surrounded by four oxygen atoms). Phospholipids are important in living things because they make up the cell membranes.

Steroids contain four rings made up of carbon. Examples of steroids are cholesterol (precursor of other steroids), testosterone (the male reproductive hormone), and estrogen (the female reproductive hormone).

Nucleic Acids

Nucleic acids are information molecules found in cells of all living things. Deoxyribonucleic acid (DNA) and ribonucleic acid (RNA) are the names of the nucleic acids found in plants and animals. Chromosomes or genes are composed mainly of DNA. DNA and RNA contain the code for amino acids in a particular order and thus tell the body what kinds of protein to make. A nucleic acid is simply a chain of nucleotides strung together and are named for the base part of the molecule.

There are only four different nucleotides in DNA named for the following four bases:

Cytosine	(abbreviated C)
Adenine	(abbreviated A)
Guanine	(abbreviated G)
Thymine	(abbreviated T)

An example of a piece of DNA molecule could be:

AAACTTGAGTCAT

These nucleotides are chemically bonded together. The most interesting thing about the DNA molecule is that it is the code for proteins. Each set of three nucleotides codes for an amino acid. In the above example AAA codes for lysine, an amino acid. The next set of three nucleotides or triplet - CTT codes for another amino acid and so on. All these amino acids are strung together (bonded chemically) to form a particular protein. The part of the DNA molecule that contains a group of nucleotides that codes for a protein is called a gene. Humans have thousands of genes and each code for a different protein.

Conclusion

The complexity of chemistry goes beyond our brief introduction here. The physical properties of the individual elements, their roles in reactions, and the hundreds of compounds are all subjects of interest. Geologists, medical doctors, pharmacists, engineers, homemakers, and farmers all depend upon knowing something of the discipline of chemistry. Fortunately, chemistry is the specialty of many researchers in our world and there are millions of articles and books which represent the results of good scientific research. Our focus in this chapter was to provide the reader with the basic building blocks for understanding the creation and evolution models. In the next chapter, we will use some of these building blocks to learn some more foundational material in the biological sciences.

APPENDIX C

Essential Biology

Biology is the study of life. Within the discipline of biological sciences, there are various specialties and subspecialties. The specialty or area of study may be concerned with a type of organism (*e.g.,* snakes), a particular organism (*e.g.,* Western Plains Garter Snakes), a particular part of an organism (*e.g.,* the tongue of the Western Plains Garter Snake), a particular structure (*e.g.,* the epithelial cells of the tongue of the Western Plains Garter Snake). Other areas of study may be interested in relationships among organisms and the physical world (*e.g.,* ecology). Some important disciplines or fields within the biological sciences include:

Discipline	Area of Study
Botany	Plants
Zoology	Animals
Ethology	Animal Behavior
Ecology	Organisms & Environment
Entomology	Insects
Herpetology	Reptiles
Mammalogy	Mammals
Ichthyology	Fish
Ornithology	Birds
Microbiology	Bacteria, Viruses
Genetics	Inheritance
Biochemistry	Chemistry of Living Things

All of these areas of study are very complex and literally hundreds of thousands of articles and books can be found that describe and explain these areas. Regardless of the organism or the part of an organism, all this research is dealing with one or more of the basic concepts of biology: (1) the organization of life, (2) the classification of the organisms, (3) structure and function of the organism, or (4) the mechanism of inheritance. Many, if not most, of the arguments for evolution and creation center around conclusions drawn from the principles developed in these areas. Therefore, this chapter will provide the basics of biology in order for the reader to better understand these arguments.

Living Things

The following characteristics define all living things:

- All living things contains DNA.
- Cells are the basic unit of all life.
- Living things are adaptable to a changing environment.

Living things are found in the earth's **biosphere**. The biosphere includes both the earth's crust and the sky. Organisms, whether plants or animals, can be studied as individuals or in a group. A group of the same kind of plant or animal is referred to as a **population**. Organisms are usually found together, coexisting with other organisms in a small area, such as a pond or woods. This is referred to as a **community**. The community of organisms, along with its environment (climate, geology, etc.) form an **ecosystem**. Large areas of similar ecosystems are called **biomes**. Examples of biomes are deserts, deciduous forests, or tropical rain forests.

Classification of Living Things

Scientists organize and name living things in an orderly manner. Taxonomy is the process of classifying organisms in established categories. Biologists use the Linnean system of classification which was invented and published by Carl von Linne (also known as Linnaeus) in 1735.

In the Linnean system, the **kingdom** category is the highest division of living things and subsequent categories get progressively subordinate. There are five kingdoms. Within each of these kingdoms, every living thing can be classified:

- **Prokaryotes** are simple (no nucleus), one celled bacteria and cyanobacteria (blue green algae). They have cell walls.

- **Protista** are one celled organisms. They are photosynthetic or heterotrophic (eat other organisms). Examples include protozoa, amoebae, all algae, diatoms, slime and water molds.

- **Fungi** contain filaments and cell walls. They include the molds, mushrooms, yeasts.

- **Animalia** are multicellular and have cell membranes instead of cell walls. Some examples range from insects to mammals.

- **Plants** are multi-cellular and contain a cell wall and chlorophyll for photosynthesis. Some examples are mosses, ferns, trees, flowers, and grasses.

- **Viruses** can be placed in a separate kingdom or in with the prokaryotes. Viruses are not composed of cells, but only DNA or RNA, and a protein coat.

The kingdoms are subdivided into **phyla**. Within the phyla, there are **classes** of organisms. Each class contains various **orders** and within the orders there is another subdivision called **family**. It is within the families that the basic units of classification are found — the **genus** and the **species**.

A genus is a taxonomic grouping of many species that are similar. There are two specific names, a genus name and a species name that are used to classify every organism, plant or animal. This name is unique and not given to another plant or animal. The species is a group of animals or plants that are very similar, almost identical. They usually look alike and have very similar genes. By definition, they can interbreed and reproduce offspring with one another but not with different species. For example, dogs can interbreed with other varieties of dogs. However, dogs can't interbreed with cats. The following Table C-1 illustrates the classification system as it applies to three examples:

Taxonomic Classification	Human	Sweetcorn	Butterfly
Kingdom	Animalia	Plantae	Animalia
Phylum/division	Chordata	Anthophyta	Arthropoda
Class	Mammalia	Monocotyledons	Insecta
Order	Primate	Commelinales	Lepidoptera
Family	Hominidae	Poaceae	Danaidae
Genus	Homo	Zea	Danaus
Species	sapiens	mays	plexippus

Table C-1. Example Comparative Classification Between the species Human, Sweetcorn, and a Butterfly.

Further subdivision within orders provides a refined classification. Humans are differentiated from monkeys and apes in the following way:

Within the order Primates, there are two suborders:

(1) Prosimii - lemurs and tree shrews
(2) Anthropoidea - monkeys, apes and humans

Within the suborder Anthropoidea, there are three superfamilies:

(1) Ceboidea - new world monkeys (squirrel monkeys and spider monkeys)
(2) Cercopithecoidea - old world monkeys (baboons and rhesus monkeys)
(3) Hominoidea - apes and humans

Within the superfamily Hominoidea, there are three families:

(1) Hylobalidae - gibbons
(2) Pongidae - chimpanzees, gorillas, and orangutans
(3) Hominidae - humans

The classification system is a good way to categorize organisms for the purposes of study. Humans have many characteristics in common with monkeys and apes; therefore, we are classified as members of the same taxonomic superfamily. Remember, the taxonomic system is man-made and shouldn't be used to make inferences about ancestry. The existence of a species (living or extinct) is not dependent upon evolution and is certainly not dependent upon a system which implies a relationship of one organism to another. Taxonomy does not and will never be a system for inferring or conferring relationships between species.

Structure of Living Things

Complex organisms, regardless of their classification, are made of the same basic four-level structural hierarchy:

Organ system such as the circulatory system (heart, lungs, blood vessels).
Organs such as the lung, the heart, the kidney, etc.
Tissues such as muscle, nervous, connective, and epithelial
Cells such as muscle cells, bone cells

Cells and Their Composition

Cells are the basic unit or building block of all living things except viruses. Cells are really where all the action takes place. Cells are where the food is burned and energy is released for use. Most disease processes occur at the cellular level. When evolutionary biologists talk about evolution of animals and man, the evolution they are talking about is thought by them to begin at the cellular level. It is important to have a basic understanding of the activities occurring at this level.

Within the cells there are **cell organelles** such as mitochondria and the cell nucleus. Certain molecules such as glycogen, glucose, and fat are also found inside the cell. The following are descriptions of some of the major cell organelles which comprise cells and serve various functions. An electron micrograph picture of the inside of cell is shown in Figure C-1.

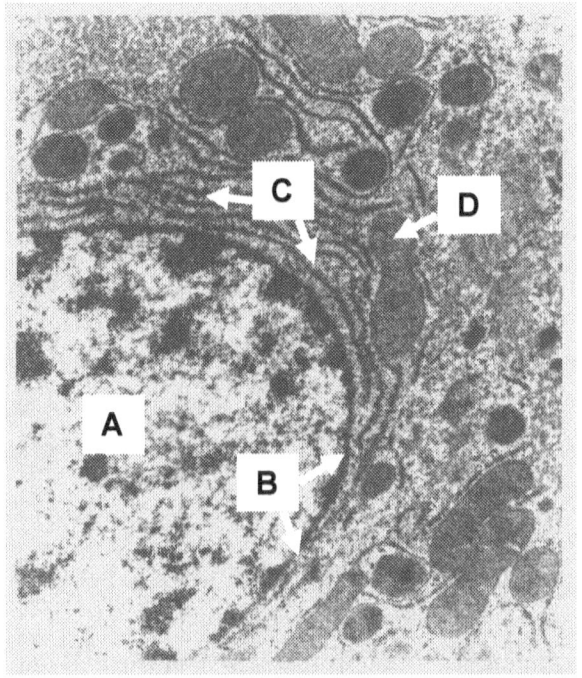

Figure C-1. Electron Micrograph of a Cell.
A - Nucleus (Containg the DNA); B - Nuclear Membrane;
C - Endoplasmic Reticulum; D - Mitochondria

The **cell membrane** surrounds and holds the cell together. It is selectively permeable; that is, it allows only some substances to pass through. The membrane structure is made up of a double layer of phospholipid (fat) molecules.

The **endoplasmic reticulum** (ER) is a complex network of membranes for transport of molecules in the cell. The rough endoplasmic reticulum has ribosomes on it. The smooth endoplasmic reticulum is where lipid synthesis occurs.

Ribosomes are composed of protein and nucleic acids. They are the protein factories of the cell. The ribosome manufactures proteins from the components of amino acids. The ribosomes are located in the endoplasmic reticulum and in the cytoplasm (the substance between the cell membrane and the nuclear membrane).

The **Golgi apparatus** is found near the nucleus and is composed of a stack of about six membranous sacs. It packages the proteins made in the ER by adding on sugars. It then sends the protein to the exterior of the cell wrapped in a portion of the Golgi membrane.

Mitochondria are the powerplants of the cell. Energy in the form of adenosine tri-phosphate is made here by oxidizing (burning) glucose.

Lysosomes are sacs of powerful enzymes that digest anything with which it comes in contact. It is the cell recycling center. For example, it decreases body tissues such as the uterus after delivery of a baby.

A **centrosome** is composed of two centrioles which are hollow cylinders made up of protein. They are involved in cell division and help chromosomes divide.

The **cell nucleus** is found in the center of the cell and contains the genes of the cells. It is like the brain of the cell since it directs the cell's activities. It contains the nuclear membrane, nucleolus and chromatin.

The **nuclear membrane** is a semipermeable double membrane, like the cell membrane, and it prevents the genes or DNA strands from leaving the nucleus and keeps other molecules out. The nucleolus is a small sphere of RNA and protein. It synthesizes ribosomes. The ribosomes then leave the nucleus through pores in the nucleus. There can be more than one nucleolus in the nucleus.

Chromatin is the DNA in the uncoiled state like a bunch of spaghetti or a mass of yarn. DNA in the coiled state is called a chromosome. It contains information (genes) on how to make a protein. There are 46 separate chromosomes in each nucleus of every cell in the human body (except the reproductive cells).

Cell Division

Chromosomes are composed of DNA in tightly wound strands. Each cell in an animal or plant contains the same number of chromosomes except for the gametes (eggs and sperm). Table C-2 gives some examples of chromosome number:

Organism	Chromosome Number
Humans	46 chromosomes (23 pairs)
Chicken	78 chromosomes (39 pairs)
Goldfish	94 chromosomes (47 pairs)
Corn	20 chromosomes (10 pairs)
Cat	38 chromosomes (19 pairs)
Mosquito	6 chromosomes (3 pairs)
Dog	78 chromosomes (39 pairs)

Table C-2. Examples of Chromosome Numbers for Seven Organisms

Homologous pairs of chromosomes are chromosomes that are the same size and carry identical functional genes such as the hair color gene. For example, the eye color gene can be for blue color on both chromosomes or blue color on one chromosome and brown on the other chromosome. These varieties in genes are called **alleles**. In other words brown, blue, and green are examples of the eye color allele.

For humans there are 46 chromosomes or 23 pairs. Chromosomes have a number assigned to them; 1 through 23. The number 23 chromosomes are special and are called sex chromosomes. They are referred to as the X and Y chromosome. All 23 pairs of chromosomes are found in every cell in the body, but the genes of all the chromosomes are not in use in every cell. For example the eye color gene is not in use in the liver cells but only in the eye cells. An exception to the chromosome number occurs in gamete cells (See Meiosis below).

Mitosis is the process of nuclear and chromosome division. One original cell divides into two new cells which are genetically identical. In humans the original cell has 46 chromosomes (23 pairs) and the two new cells have 46 chromosomes (23 pairs) each.

Meiosis is the process by which gametes (sperm and egg) are formed. The chromosome number is halved in animals. For example, in humans 46 chromosomes are reduced to 23. This occurs so that when the egg cell and sperm cell combine at fertilization, there is a total of 46 chromosomes. Meiosis only occurs in the ovary of the female and in the testes of the male.

Genetics

Genetics is another subset or specialty of biology. It is the study of how characteristics or traits, such as hair color, are passed to the offspring of living things. A trait is a certain characteristic in any living organism such as hair color or it can be a particular protein in the body such as insulin. Traits are controlled by a gene on a chromosome and traits are inherited. Examples of inherited traits:

- Eye color
- Blood proteins such as hemoglobin
- Skin color
- Blood type
- Instincts such as thirst or hunger

A **gene** is a region on the chromosome comprised of a series of nucleotides. Most genes code for proteins. It is the protein that is responsible for the trait. For example, the eye color gene codes for a protein that causes the eye to have a certain color. When studying genetics, the gene can be symbolized by a letter of the alphabet. For example, a pea seed can have the gene which makes it a yellow seed or a green seed. A capital Y is used to represent the yellow seed because it is the dominant seed color and a small y is used to represent a green seed, since it is recessive.

Y = yellow seed   y = green seed

An **allele** is an alternative form of the same gene. In this example, there is a yellow color allele and a green color allele in the pea plant (Y or y). It's possible for organisms to possess both alternatives, even though it only exhibits one of them. A **genotype** is the actual genes that the plant or animal cell contains. There are always two alleles because the chromosomes that are comprised of these genes are paired. One allele (gene) is on one of the chromosomes and the other allele is on the other. For example, a yellow pea seed could be YY (both alleles for yellow) or Yy (one yellow and one green). However, since Y is dominant, the plant will have yellow seeds. On the other hand, the **phenotype** is the expression of genes in the animal or plant. For example in the pea plant, the phenotype would be a yellow pea seed or a green pea seed. It is the trait of the plant or animal that is expressed or seen in the plant or animal.

Mendelian Genetics

Mendelian genetics is fairly simple and straightforward genetics that was first discovered and explained by Gregor Mendel. Gregor Mendel was a Catholic priest in Czechoslovakia in 1856-1864. He was hired to breed fruit plants such as peas because he was also a biologist and mathematician. He decided to study pea plants in detail. He knew something controlled traits in pea plants and he called them factors which we call genes today. He found each trait was controlled by two factors which we know are the two alleles of the gene. He used math to analyze his results of his breeding of many pea plants. He discovered that one allele was dominant and one was recessive for each of seven traits that he studied in the pea plant. Examples of traits that Mendel studied include:

T = tall pea plant   **t** = short pea plant
Tt are tall plants called heterozygous (both genes present).
TT are also tall plants called homozygous dominant
(only the dominant genes are present).
tt are short plants called homozygous recessive
(only the recessive genes are present.

Y = yellow seed   **y** = green seed
Yy have yellow seed called heterozygous
YY have yellow seeds called homozygous dominant.
yy have green seeds called homozygous recessive.

Modern Genetics

There are some exceptions discovered since Mendel. Although not contradictions, modern genetics studies traits that do not follow the simple principles of Mendelian genetics. There are several illustrations of modern genetic principles, including polygenic inheritance, multiple alleles, and co-dominance.

Polygenic Inheritance. This is where two or more genes control a trait. Some examples are skin color, eye color, and height. More than one gene controls height and, as a result, there are not just two different heights in humans which is what there would be if height was controlled by one gene. There is a wide range of heights which tells biologists that there is more than one gene for height. It is known that tall genes are recessive to short genes for height. Eye color is determined by the amount of melanin in the eye. The more genes one has that have the allele for melanin, the darker the eye color which ranges from dark brown (most melanin) to light brown to hazel to green to blue (which hardly contains any melanin). Red eyes do not contain any melanin at all. Skin color is a similar situation. Light skin contains hardly any melanin because there are few genes coding for melanin in those cells. Dark skin has more genes that code for melanin.

Multiple Alleles. This is where more than two alleles are possible for a trait. For example:

- Blood type A allele produces A protein
- Blood type B allele produces B protein
- Blood type O allele produces no protein.

A and B are both dominant alleles and O is recessive. Therefore, there are many blood types or blood phenotypes possible. Type O blood can only be OO genotype since O allele is recessive. Type AB blood is AB genotype because A and B allele are co-dominant. Type A blood can be AA or AO genotype. Type B blood can be BB or BO genotype.

Co-dominance occurs where there is no dominant or recessive allele. An example is the four o'clock flower with only two alleles:

R = red   **r** = white

The phenotypes are:

RR = red   **Rr** = pink   **r** = white.

Conclusion

After seeing the beauty, complexity, and high degree of organization in the biological world, it's difficult to believe that anyone would try to explain life as a product of random chance events. However, that's exactly what the evolutionist does. It seems much more logical and reasonable to see the hand of a designer or Creator. With an understanding of the principles of biology, the creation model should be better understood and appreciated.

APPENDIX D

Paleontology, Fossils, and Dinosaurs

Fossils are the preserved remains of things once living. The remains may occur in a variety of forms, from impressions to mummies. Paleontology is the scientific discipline that studies fossils. Paleontology is similar to biology with one major difference. Biology, and it associated subspecialties, is concerned with the study of living things. The paleontological sciences are focused on things that were once living, but are now considered ancient and are, for the most part, fossilized. Some of the specialties in paleontology include:

Discipline	Area of Study
Paleobotany	Ancient and fossilized plants
Paleozoology	Ancient and fossilized animals
Paleoanthropology	Primitive man

Paleontological Methods

Paleontologists employ the same scientific approach to study that biologists use, but have some materials and methods that are applied specifically for the location, extraction, analysis, and handling of the various specimens and artifacts of interest. Studying these ancient forms can be helpful in morphological comparisons between modern and ancient, understanding conditions that may have led to an organism's demise, and appreciating the diversity of life that has lived upon the earth. Paleontology is an important discipline for assessing the arguments surrounding the evolution and creation controversy.

The discipline of paleontology uses some accepted techniques for ascertaining some of the unknowns, based on the known living world.

171

For example, the insertion points for muscles to bone can be determined by examination of the fossilized bone specimen. The way bones are connected (articulated) can also be determined this way as well. Brain size can be determined fairly accurately by assessing the cranial volume of a fossilized skull.

The classification of fossils is done using the same classification system used for living things — the Linnean taxonomic system. Often times with fossils, little is known of the soft-tissue anatomy, such as internal organ structure, muscle, and skin characteristics (*e.g.,* presence of hair, fur, scales, color, thickness, etc.) Without these soft-tissue indices, it is difficult (and even impossible) to classify some things as one species or another. Therefore, fossils are classified by paleontologists largely based on amount of variation. Much variation between a fossil and a supposed living species would result in the fossil being classified as a different (possibly extinct) species.

Fossils

We talked some about the fossils in the creation chapter and the evolution chapter. As we said, the fossil record is a friend to the creation model. It explains very nicely that there was a catastrophic event in history (the flood of Genesis) which devastated life on the earth at that time.

Fossils are natural representations of organisms that have died. Any organism that dies is a candidate to become a fossil, however, many do not become fossils for a variety of reasons. One of the primary requirements to become a fossil is for the organism to be rapidly covered in order to prevent decay or consumption by other organisms. The sediment produced in floods is a good example of how this occurs. In fact, the flood of Genesis undoubtedly produced conditions which would entrap organisms in a way to promote fossilization. This mechanism is based on models developed from data from observed floods. The sediment produced would have covered organisms living at the time of the flood (of Genesis) and the fossilization process would then take place.

It's not surprising that ninety-nine percent of the world's fossils are found in sedimentary rocks formed by receding flood water. Given the nature of the catastrophe of the Genesis flood, it is also not surprising that there is an abundance of marine (or ocean) fossils found at high elevations and even in mountains. There must have been a tremendous upheaval during the flood event and we know that all the dry land was covered. When we were in Turkey, in the 1980s, we spent time climbing in the Taurus Mountains, north of Tarsus (the Apostle Paul's home town). Once, we were climbing at about 9,000 feet above sea level and found an abundance of fossil marine organisms including many types of shells and coral (Figures D-1 and D-2).

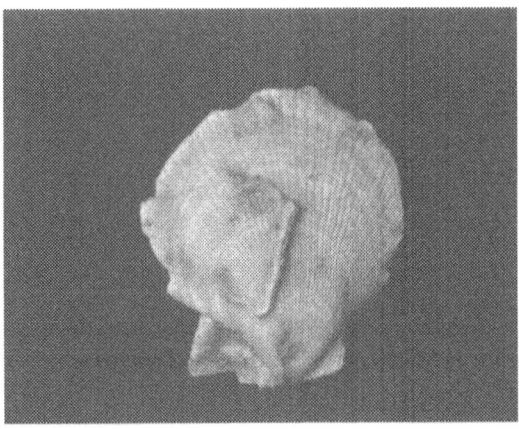

Figure D-1. Fossil scallop shell approximately two inches in diameter; both halves intact.

These were interesting discoveries for us, but there have been many such fossil finds in mountains around the world. A very early discovery was made in the 16th century by Bernard Palissy who was a collector and student of fossil shells and fish. He wrote "I have drawn a number of pictures of the petrified shells that can be found by the thousands in the Ardennes Mountains, and not just shells, but fish...I have found more kinds of petrified fish, or the shells thereof, than I have of modern kinds now living in the sea."[75] Something definitely happened to trap this large variety of organisms.

Figure D-2. Fossil coral 3.5 inches long.

How Fossils are Formed

There are a variety of ways once living organisms are preserved as fossils. The following are brief descriptions of some of these common processes:

Freezing is similar to putting food in the freezer. With this process, an organism is preserved from the decay process. Decay is caused by the invasion by microbial organisms that feed from the dead organism. While alive, an organism is protected by intact skin and naturally occurring anti-microbial agents in its body. However, once dead, the organism is vulnerable to this decay. If the organism is encased in ice, this provides an added benefit of protecting it from

---

[75] Yvette Gayrard-Valy, *Fossils Evidence of Vanished Worlds* (New York: Henry N. Abrams, Inc., 1994), p. 35.

other consumers that may have otherwise have eaten its flesh. Some good examples of frozen fossils are the mammoths frozen in Siberia. Some very interesting examples of human fossils preserved this way are the Inca sacrifices found entirely preserved high in the Andes Mountains in Argentina. These 500 year old remains provide a detailed representation of the once living human.

Drying or desiccation provides another fossil preservation method that protects the organism from extreme decay. Because of the rapid drying in an extremely arid environment, such as a desert, microbial invaders are reduced and eventually the organism is preserved. An organism that may have met it demise in a desert, then covered over with sand, may be preserved for centuries in a desiccated state. Unlike the frozen fossils, these have lost some of the original characteristics, such as blood, due to the dehydration. However, much of the structure is still very well represented, even on the cellular level. The drying mechanism works with freezing to enhance preservation. Sometimes chemical preservation, used in mummification provides an early preservation as the drying process begins.

Petrification occurs when the original material of the organism is replaced by other mineral material. This may be a complete replacement (if the specimen is old enough) or a partial replacement. This fossilization process requires the exchange of the original living chemicals with surrounding chemicals. For example, bone trapped in a limestone (calcium carbonate) matrix will gradually become stone. This explains why the originally white bone becomes dark and discolored — it has taken on the color of the mineral within which it has become fossilized. The flesh of the organism has been consumed by other organisms, but the hard bone remained to become fossilized. So-called petrified wood is preserved in this way as well.

Carbon film preserves an impression of soft-bodied animals and plants. The original composition of the organism is changed by bacteria, chemicals, heat, and pressure. What remains is a thin carbon film which provides a detailed representation of the original structure. An example is a leaf "impression" in rock.

Casts and Molds are formed when an organism is trapped in sediment and in time it dissolves leaving an empty space or cast that represents an exact representation of the. reverse of the outside of the organism. After more time passes, this cast may refill with a replacement material and this forms a mold of the original organism's shape. Sometimes the inside of the organism is filled with sediment that eventually hardens. This internal mold is called a steinkern. Shells are commonly fossilized as casts and molds.

Figures D-3 and D-4 are two more examples of common marine fossils.

Figure D-3. Fossil fish 1.75 inches long from the Green River formation, Wyoming.

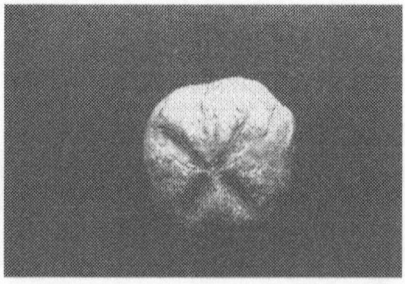

Figure D-4. Fossil echinoderm (e.g. sand dollar, sea urchin) 2 inches in diameter.

Dinosaurs

Dinosaurs are always fascinating organisms to think about and study. They have a place in the study of modern biology as a group of extinct reptiles. Of all the fossilized animals, the dinosaur's popularity is unsurpassed and for that reason, we will showcase them as a product of paleontological work. The dinosaur was a member of the group of animals called vertebrates in the phylum Chordata. These animals had back bones composed of small bones known as vertebrae. Vertebrates also have a skull which encloses the brain.

Within the vertebrates there are many groupings of animals known as classes. For example, there is the mammal class, the fish class, the bird class, and the reptile class. The dinosaurs belong to the reptile class even though they are extinct. They were a larger reptile, more like a lizard than a snake or turtle. They were created along with the other animals and man, so they lived with man before the flood. If they were nonextinct before the flood, Noah certainly took representatives of them on the ark. Some may have lived for a time after the flood then became extinct. Evolutionary scientists place dinosaurs in the Triassic, Jurassic, and Cretaceous periods up to 235 million years ago.

For some reason, the dinosaurs disappeared (*i.e.,* they became extinct). Scientists think the extinction might be due to the climate becoming colder and more unstable. The reason for their extinction fits very well into the creation/flood model. After the flood, the climate was possibly much colder and there may have been less vegetation on the earth for the herbivorous dinosaurs to feed on. They may not have adapted well to temperature change, had less food to eat, and thus died and became extinct.

Dinosaurs were generally land dwellers and walked on two legs (bipedal) or four legs (quadrupedal). In many dinosaurs the forelimbs are much shorter and slighter than the larger hindlimbs.

There are many orders of reptiles including the order Chelonia (turtles) and order Squamata (snakes and lizards). Dinosaurs are classified as belonging to one of two extinct reptile orders, the Saurischia and the Ornithischia.

Saurischia

The order Saurischia is divided into two groups, the theropods and the sauropods. The theropods were mostly carnivores and the sauropods were herbivores.

Therapods were bipedal and contain the ferocious dinosaurs. Theropods included *Ornithomimus* the ostrich dinosaur at four meters in length. There were even larger ones like *Tyrannosaurus* (12 meters long, seven tons, and 15 cm teeth — the largest flesh eating animal to ever live), *Tarbosaurus* (12 meters long) and *Megalosaurus*. These carnivores ate other herbivorous dinosaurs.

The **sauropods**, on the other hand, had small heads, long necks and very large bodies. Most were quadrupedal herbivores. The famous ones include the *Brontosaurus* (or *Apatosaurus*) at 20 meters long and 30 tons, *Diplodocus* at 26 meters long and 10 tons and *Brachiosaurus* at 23 meters long and 80 tons.

Ornithischia

The Ornithischia order includes four groups:

The **ornithopods** were herbivores and some were bipedal. They included the duck billed, parrot, and dome headed dinosaurs.

The other three groups of Ornithischia were quadrupedal, had armor for defense, and did not run fast. The **horned dinosaurs** such as *Triceratops* at nine meters in length and 5.4 tons. The **plated dinosaurs** which had thick plates that stuck out like spines all over their body; for example *Stegosaurus* at nine meters long. The

armored dinosaurs with bony chunks and spikes on them such as the *Ankylosaurus* at 10 meters in length.

Conclusion

Paleontology provides a scientific and structured methodology for understanding the origin and diversity of life. By excavating and extricating remains of organisms that once lived on the earth, many questions can be answered.

Like other disciplines, there is a lot of room for subjectivity on the part of the scientist. For this reason, care must be taken to weigh the interpretations of the paleontologists as they uncover the past. What constitutes enough variation to classify a fossil as distinct from other fossils and distinct from living species remains a debate among paleontologists. Consider the variation seen among the approximately 150 varieties of dog, yet they are all one species — *Canis familiaris*. It has been suggested that a similar amount of variation is seen in many fossils that have been labeled as a distinct species, partly for the benefit of the discoverer. These capriciously classified specimens can cause confusion in understanding the true meaning of the find.

APPENDIX E

Some Creation Related Scriptures

Selected Old Testament References to the Creation

Gen. 1:1 In the beginning God created the heavens and the earth.

Gen. 1:7 So God made the expanse and separated the water under the expanse from the water above it. And it was so.

Gen. 1:16 God made two great lights —the greater light to govern the day and the lesser light to govern the night. He also made the stars.

Gen. 1:21 So God created the great creatures of the sea and every living and moving thing with which the water teems, according to their kinds, and every winged bird according to its kind. And God saw that it was good.

Gen. 1:25 God made the wild animals according to their kinds, the livestock according to their kinds, and all the creatures that move along the ground according to their kinds. And God saw that it was good.

Gen. 1:27 So God created man in his own image, in the image of God he created him; male and female he created them.

Gen. 1:31 God saw all that he had made, and it was very good. And there was evening, and there was morning —the sixth day.

Gen. 2:3 And God blessed the seventh day and made it holy, because on it he rested from all the work of creating that he had done.

Gen. 2:4 This is the account of the heavens and the earth when they were created. When the LORD God made the earth and the heavens—

Gen. 2:9 And the LORD God made all kinds of trees grow out of the ground —trees that were pleasing to the eye and good for food. In the middle of the garden were the tree of life and the tree of the knowledge of good and evil.

Gen. 2:22 Then the LORD God made a woman from the rib he had taken out of the man, and he brought her to the man.

Gen. 5:1 This is the written account of Adam's line. When God created man, he made him in the likeness of God.

Gen. 5:2 He created them male and female and blessed them. And when they were created, he called them "man."

Exod. 20:11 For in six days the LORD made the heavens and the earth, the sea, and all that is in them, but he rested on the seventh day. Therefore the LORD blessed the Sabbath day and made it holy.

Ps. 95:5 The sea is his, for he made it, and his hands formed the dry land.

Ps. 139:13 For you created my inmost being; you knit me together in my mother's womb.

Eccl. 3:11 He has made everything beautiful in its time. He has also set eternity in the hearts of men; yet they cannot fathom what God has done from beginning to end.

Isa. 37:16 "O LORD Almighty, God of Israel, enthroned between the cherubim, you alone are God over all the kingdoms of the earth. You have made heaven and earth.

Isa. 40:26 Lift your eyes and look to the heavens: Who created all these? He who brings out the starry host one by one, and calls them each by name. Because of his great power and mighty strength, not one of them is missing.

Isa. 40:28 Do you not know? Have you not heard? The LORD is the everlasting God, the Creator of the ends of the earth. He will not grow tired or weary, and his understanding no one can fathom.

Isa. 45:12 It is I who made the earth and created mankind upon it. My own hands stretched out the heavens; I marshaled their starry hosts.

Isa. 45:18 For this is what the LORD says— he who created the heavens, he is God; he who fashioned and made the earth, he founded it; he did not create it to be empty, but formed it to be inhabited— he says: "I am the LORD, and there is no other.

Selected New Testament References to the Creation

Acts 17:24 "The God who made the world and everything in it is the Lord of heaven and earth and does not live in temples built by hands.

Rom. 1:20 For since the creation of the world God's invisible qualities —his eternal power and divine nature —have been clearly seen, being understood from what has been made, so that men are without excuse.

Col. 1:16 For by him all things were created: things in heaven and on earth, visible and invisible, whether thrones or powers or rulers or authorities; all things were created by him and for him.

Rev. 4:11 "You are worthy, our Lord and God, to receive glory and honor and power, for you created all things, and by your will they were created and have their being."

APPENDIX F

The Plan of Salvation

God's desire is for us to have eternal life with him. That's why we were created. We can have that eternal life beginning now.

> For God so loved the world that he gave his one and only Son, that whoever believes in him shall not perish but have eternal life.
> (John 3:16)

> The thief comes only to steal and kill and destroy; I have come that they may have life, and have it to the full.
> (John 10:10)

> Therefore, since we have been justified through faith, we have peace with God through our Lord Jesus Christ
> (Rom. 5:1)

The problem started with the very first man's rebellion. Having eternal life is not automatic. You have a different nature, and you are separated from Him and Life by your own sin.

> For all have sinned and fall short of the glory of God
> (Rom. 3:23)

That sin leads to judgement and death. This really means eternal separation from God in Hell. However, God provided the way for the fellowship to be restored immediately and insure our eternal salvation to be with Him in Heaven. This was accomplished by sending His Son Jesus to pay for our sins as He promised He would.

For the wages of sin is death, but the gift of God is eternal life
in Christ Jesus our Lord
(Rom. 6:23)

But God demonstrates his own love for us in this: While we
were still sinners, Christ died for us.
(Rom. 5:8)

God's way is the only way:

Jesus answered, "I am the way and the truth and the life. No one
comes to the Father except through me
(John 14:6)

For it is by grace you have been saved, through faith —and this
not from yourselves, it is the gift of God— not by works, so that
no one can boast
(Eph. 2:8-9)

What can you do?

Yet to all who received him, to those who believed in his name,
he gave the right to become children of God
(John 1:12)

Everyone who calls on the name of the Lord will be saved
(Rom. 10:13)

Pray and ask God to save you:

God, I know I am a sinner. I believe Jesus died for my sins. Right now I repent and turn from my sins. I open the door of my heart and receive Jesus as my personal Savior. I commit myself to follow you. Thank you for saving me.

Amen.

The Promise

Your commitment must be real. Merely saying a prayer, or walking down an aisle at church isn't enough. Confessing your sins is a start, but unless you are willing to yield your life in full submission to God, then you aren't truly seeking God. You must be repentant in your heart. This doesn't mean you have to "clean up your act" before you can be saved. Only God can cleanse you. It means that you must realize that there is nothing you can do to make yourself acceptable to God — but God can and will make a change in you.

That if you confess with your mouth, "Jesus is Lord," and believe in your heart that God raised him from the dead, you will be saved. For it is with your heart that you believe and are justified, and it is with your mouth that you confess and are saved (Rom. 10:9-10)

Therefore, if anyone is in Christ, he is a new creation; the old has gone, the new has come!
(2Cor. 5:17)

Jesus said: Here I am! I stand at the door and knock. If anyone hears my voice and opens the door, I will come in and eat with him, and he with me.
(Rev. 3:20)

The Lord is not slow in keeping His promise, as some understand slowness. He is patient with you, not wanting anyone to perish, but everyone come to repentance
(2 Peter 3:9)

Index

Sedimentation, 35
Simanthropus pekinensis, 109
Simon and Garfunkel, 136
Spallanzani, Lazarro, 34, 85
Species, v, 19, 20, 24, 31, 32, 35,
  37, 38, 39, 40, 41, 42, 43, 44, 48,
  49, 55, 58, 63, 64, 65, 66, 73, 74,
  80, 82, 96, 98, 99, 100, 101, 107,
  112, 113, 114, 116, 117, 118, 119,
  120, 134, 136, 160, 161, 172, 179
Spontaneous generation, 33, 34,
  82, 85, 86
Stegosaurus, 178

T

Tarbosaurus, 178
Taung Child, 114
Taxonomic system, 64, 161, 172
The Genesis Flood, 51, 63, 138
The Origin of Species, 31, 38, 42
Theory, 4, 21, 22, 23, 24, 28, 32, 33,
  34, 35, 37, 39, 40, 41, 42, 44, 45,
  63, 70, 78, 79, 80, 86, 87, 94, 98,
  100, 103, 119, 120, 121, 132, 134,
  138, 142
Theropods, 178
Transitional forms, 40, 42, 43, 45,
  48, 80, 82, 96, 98, 99, 100, 101,
  102, 105, 112, 116, 119, 120
Trilobite, 126

U

Uniformitarian, 72, 131
Uniformitarianism, 35, 36, 123
University of Chicago, 24, 35, 45,
  87
University of North Carolina, 104
Uranium, 127, 129, 130, 149

V

Van Leeuwenhoek, Anton, 34
Vapor canopy, 51, 67, 68, 69, 70,
  71, 74, 130
Variation, v, 19, 31, 32, 37, 40, 41,
  42, 47, 48, 55, 65, 66, 67, 73, 74,
  80, 82, 99, 100, 106, 112, 113,
  118, 126, 130, 134, 172, 179
Virchow, Rudolf, 34, 86
Viruses, 79, 157, 159, 162
Von Linne, Carl, 55, 63, 64, 159

W

Walker, Alan, 117
White, Tim, 96, 113

Y

Young earth, 23, 35, 36, 82, 121,
  123

Robert and Elizabeth Ridlon

About the Authors

Robert W. Ridlon, Jr. is a systems development consultant and has been an adjunct faculty member at Southwestern Illinois College, Illinois since 1991 teaching in the Computer Information Systems Department. Mr. Ridlon earned a Bachelor of Arts degree in Biological Sciences from Indiana University in 1978 and a Master of Science degree in Information Resource Management from the Air Force Institute of Technology in 1989. His master's thesis was in the area of organizational behavior. Mr. Ridlon has published in the areas of biology and information systems. Mr. Ridlon has been an ordained Southern Baptist deacon since 1991.

Elizabeth Ridlon is an adjunct faculty member at Southwestern Illinois College where she teaches in the Biological Sciences Department. Previously, she taught biology at the University of Maryland, European Division. Mrs. Ridlon earned a Bachelor of Science degree in Microbiology from Indiana University, a Master of Arts degree in Biology from the University of Nebraska, and a Secondary Teachers Certificate from Southern Illinois University at Edwardsville. Her master's thesis was a study of the seasonal changes in the blood biochemistry of the Western Plains Garter Snake. Mrs. Ridlon is also a Precept Bible Study teacher.

Robert and Elizabeth have given numerous creation science talks to home school groups and conferences, church classes and groups, and Campus Crusade for Christ. The Ridlon's previous books on creation are titled *Understanding the Origin and Diversity of Life* and *Creation Science Made Easy*. In addition, Mr. Ridlon has spoken on creation and evolution at Indiana University, University of Illinois, Lewis and Clark College, and on Christian radio. The Ridlon's have visited and explored four continents that included the countries of Israel, Egypt, Greece, Italy, England, Germany, Austria, Switzerland,

Belgium, Luxembourg, Netherlands, France, Turkey, Canada, Mexico, Spain, and Morocco.

www.ingramcontent.com/pod-product-compliance
Lightning Source LLC
Chambersburg PA
CBHW032003170526
45157CB00002B/517